SKY
WARRIORS

SKY WARRIORS

CLASSIC AIR WAR BATTLES

ALFRED PRICE

CASSELL

Cassell Military Classics

Cassell
Wellington House, 125 Strand
London WC2R 0BB

First published by Arms and Armour 1994
This Cassell Military Classics edition 1998
Reprinted 1999, 2000

British Library Cataloguing in Publication Data
A catalogue record for this book is available from
the British Library

ISBN 0–304–35130–X

Edited and designed by
Roger Chesneau/DAG Publications Ltd

Printed and bound in Great Britain by
Cox & Wyman Ltd., Reading, Berks.

CONTENTS

INTRODUCTION

AS IN THE CASE of the companion volume, *Sky Battles*, this book describes in detail several air actions fought during the past eight decades, and uses these to illustrate the multi-faceted nature of air warfare.

The narrative opens in September 1917, when the First World War was in its fourth year and in Europe the opposing armies were locked in the deadly stalemate of trench warfare. Chapter 1, 'By Zeppelin to Africa' describes the epic attempt by Zeppelin *L59* to transport supplies to the German troops cut off in East Africa. In the event, the venture was unsuccessful, but the failure was not due to any shortcoming on the part of the airship or her intrepid crew. Had the Zeppelin arrived at her destination there can be no doubt that the flight would have been hailed as a stunning propaganda *coup*, and quoted later as a shining example of what a single, resolutely handled aircraft could achieve. As it was, the 95-hour flight by the airship established a record for the length of a combat flying mission that has never been broken.

The narrative then moves on to the Second World War. During the 1930s two new areas of technology combined to bring about a revolution in air defence and air fighting tactics: radar and radio telephony. The former enabled ground controllers to gain early warning of the approach of enemy raiding forces and send off large numbers of single-seat fighters in time to meet the threat. The latter made it possible to assemble fighters in unprecedented concentrations and direct them into action against the enemy force. The Battle of Britain in the summer of 1940 was the first occasion when both of these factors came into play in a large-scale air battle. It was also the first decisive military action to be fought solely in the air, without no serious involvement from either ground or naval forces. The action described in detail in Chapter 2, 'Battle of Britain Day' was that fought over London on 15 September 1940 and commemorated each year. Nothing can, nor should, detract from the bravery of the RAF fighter pilots that went into action that day. But it was also a victory for the system of fighter control that Air Chief Marshal Dowding had painstakingly built up from first principles. The *Luftwaffe* mounted two attacks on London that day. To meet the first, Fighter Command scrambled twenty-three squadrons of Spitfires and Hurricanes and all except one made contact with the enemy. To meet the second attack Fighter Command scrambled twenty-eight squadrons, and every one of them went into action. No conceivable system of fighter control could have done better than that.

Not only was the Second World War the first conflict to see the assembly of really large forces of combat planes to deliver a co-ordinated attack, it was also the first in which an air arm could deliver a sufficiently large weight of high explosive to cause major destruction to targets. The combination of the two made the pre-emptive air strike an attractive option to any regime ruthless enough to employ it. In such an attack the transition from peace to war came like 'a bolt out of the blue', with devastating air strikes hitting simultaneously at a spread of targets. Usually these were followed by powerful armoured thrusts by ground forces. The technique had been pioneered in Poland, Belgium, Holland and Yugoslavia. Chapter 3, 'Pre-Emptive Strike' describes the largest and the most effective of them all, that on Sunday 22 June 1941 when Adolf Hitler unleashed his *Blitzkrieg* offensive against the Soviet Union. The day was certainly one for the aviation record books. In an eighteen-hour period between 3.15 a.m. and sunset the Soviet Air Force lost about 1,800 combat planes, most of them destroyed on the ground. It was by far the largest number of aircraft wrecked in a single day's fighting and it represented the most comprehensive defeat ever inflicted by one air force on another. More than 300 Soviet aircraft fell in air-to-air combat, the largest number ever shot down in a one-day period.

The advancing German troops quickly overran every one of the airfields that the *Luftwaffe* had attacked on the first day. This added to the losses already suffered by the Soviet Air Force, for any damaged or unserviceable planes that could not be flown out before their airfields were captured were lost. Moreover, the capture of the airfields enabled the *Luftwaffe* to move to bases closer to targets in the Soviet hinterland. As a means of securing air superiority, the capture of enemy airfields is a highly effective method that is sometimes omitted from such calculations.

Despite the enormous material losses suffered by the Soviet Air Force during the early days of the war, however, the effect on that service was crippling only in the short and the medium term. Very few air crew were lost when the airfields were attacked and later captured, so as new aircraft became available there was no shortage of trained personnel to put them into action. Furthermore, the Soviet Air Force was on the point of re-equipping its front-line units with more modern fighter and bomber planes; most of the aircraft lost during the initial onslaught were obsolescent types that were about to be replaced anyway.

Chapter 6, 'Hard Fight to "The Big B"' describes the first large-scale US Army Air Force attack on Berlin, on 6 March 1944. These deep-penetration attacks became a feasible proposition only when long-range escort fighters, and in particular the superb P-51B Mustang, were available in large numbers to protect the bombers. On that day the Eighth Air Force lost 69 bombers and eleven fighters; it was a heavy loss, but there were sufficient replacement crews and aircraft to fill the gaps in the ranks of the front-line units. During the action the *Luftwaffe* lost 66 fighters and 46 pilots killed or wounded, and while the losses on planes were easily made good, the crews were almost irreplaceable. Following that attack, no target in Germany was beyond the reach of American heavy bombers protected by the escorts. The *Luftwaffe* was inexorably losing control of the skies over its homeland, and would never regain it.

During the final months of the Second World War night raiders were also being escorted to and from their targets, not only by long-range Mosquito night fighters but also by special jamming aircraft carrying equipment to neutralize or spoof defending radars. These operations were the work of No 100 Group of the Royal Air Force, and they are described in Chapter 10, 'Confound and Destroy'.

As we have observed, a well-executed air attack can inflict enormous damage on an adversary. By the same token, however, an air attack that is ill-conceived or badly carried out can result in heavier losses for the attackers than for the attacked. Two clear examples of the latter are described in Chapter 7, 'The Great Marianas Turkey Shoot' and Chapter 11, 'New Year's Day Party'. In the first of these the Japanese Navy launched a massive carrier air strike against a US Navy carrier task force. Had the Japanese Navy possessed a better fighter at that time than the aging A6M5 Zeke 52, and had its air crews received a level of training comparable with those sent into action earlier in the war, the battle might have ended as a resounding victory for that service. As it was, the Japanese attack forces were massacred by defending American Hellcat fighters which, together with the ships' guns, shot down 181 of the attackers in a morning for a loss of only seven US fighters.

Chapter 11, 'New Year's Day Party' describes the *Luftwaffe* attack on Allied forward airfields in Europe on 1 January 1945, when that service attempted to repeat the success it achieved against the Soviet Air Force in June 1941. But the Allied Air Force units based in France,

Holland and Belgium were a good deal better prepared to meet such an attack than the *Luftwaffe*'s earlier victim had been. Moveover, by the beginning of 1945 the *Luftwaffe* was a good deal less able to deliver an effective blow than it had been in 1941. As a result, the attack failed at several points and the *Luftwaffe* lost rather more aircraft in the action than the Allied air forces. Even worse, the *Luftwaffe* lost 237 pilots killed, wounded, missing or taken prisoner that day and these included several experienced fighter commanders who were quite irreplaceable. Allied pilot losses during the action were minimal—probably less than twenty. The German fighter force never recovered from the blow that it suffered that day.

At the beginning of the Second World War the range of offensive weapons available to air forces was strictly limited. Often crews faced appalling risks in order to deliver their bombs on the assigned targets when, with hindsight, it is clear that those weapons had little chance of achieving a worthwhile degree of damage. For example, during its valiant efforts to stem the Germany advance through Holland, Belgium and France in May 1940, the Royal Air Force sent obsolescent Blenheims and Battles armed with 250lb general-purpose bombs to attack bridges. This weapon was insufficiently powerful to destroy a steel or masonry bridge even if several of them scored direct hits on the structure. The bombers suffered horrifying losses and, predictably, they achieved no tangible results. During the war each of the major belligerent air forces had to learn the same lesson, and in the hardest possible school of all—that of combat. The Royal Air Force went on to develop a range of specialized aerial weapons that were highly effective against their intended target. In Chapter 9, 'Bombers against the *Tirpitz*' we look at two types of specialized weapon that were employed in attacks on the mighty German battleship—the ingenious 'Johnny Walker' underwater walking mine and the 'brute force' 12,000lb 'Tallboy' bomb. To attack the *Tirpitz* was a man-size job, and certainly the 'Tallboy' was a man-size weapon with which to do it.

The Second World War also saw the introduction into service of the first air-launched guided weapons. Such a weapon can be defined as 'a missile whose path can be corrected during its travel, either automatically or by remote control, to bring it into contact with the target'. If that definition is accepted, the reader will note that it makes no mention of the medium through which the weapon passes, nor of its speed. While most guided weapons travel through the air at high speed, the first such air-launched weapon did not: it was in fact an anti-submarine homing torpedo, and its maximum speed when running underwater was 12kts. Developed in the United States, the homing torpedo was code-named the 'Mark 24 Mine' to conceal its true purpose. Its operational career is described in Chapter 4, 'Guided Missiles Make Their Mark (1)'. In the course of the Second World War these homing torpedoes were credited with the destruction of 38 German and Japanese submarines, and with inflicting damage to a further 33.

During the Second World War Germany fought two of the world's largest maritime powers, Great Britain and the United States, each of which possessed a navy that was far larger than her own. To redress this imbalance the *Luftwaffe* accorded the highest priority to the developed of specialized weapons for use against enemy shipping. Two types of radio command-guided missiles resulted, the Henschel Hs 293 glider bomb and the Ruhrstahl Fritz-X guided bomb, and their operational careers are described in Chapter 5, 'Guided Missiles Make Their Mark (2)'. During August and September 1943 these weapons scored a string of successes. Fritz-X bombs sank the Italian battleship *Roma* and inflicted severe damage on her sister ship *Italia*, the Royal Navy's HMS *Warspite* and several Allied cruisers. The Hs 293 sank a number of smaller warships. The Allies quickly discovered the secrets of the missiles' radio guidance system, however, and introduced a special type of transmitter designed to jam the guidance signals. The result was that, from the spring of 1944, the two German guided missiles achieved little.

To take the story of specialized weapons into the jet age, in Chapter 12, 'Target Hanoi', we examine the first use of laser-guided and electro-optically guided 4,000lb bombs against a target in North Vietnam. The attack described is that on the famous Paul Doumer Bridge, in May 1972. Even a score of these powerful weapons, aimed with unprecedented accuracy at vulnerable points on the structure, failed to drop a span during the first attack. But they rendered the bridge unusable to wheeled traffic, and on the next day a follow-up attack, in which these weapons were aimed into the previously weakened area, did drop one of the spans. The narrative gives an insight into the force-package type of operation, showing the fighter escort and defence-suppression techniques that had evolved in the quarter-century since the end of the Second World War. From cockpit voice-recording tapes made in the aircraft at the time of the action, the reader can get a sense of the drama of modern air combat.

Airfield runways, like bridges, are small, hard targets that are very difficult to attack effectively unless specialized weaponry is used. Yet in Chapter 13, 'The First "Black Buck"', we see a Vulcan bomber having to do the job in with a stick of twenty-one 1,000lb 'dumb', general-purpose bombs for want of anything better. For the sheer determination of the crew to succeed in the face of atrocious weather conditions and mind-blowing distances, that raid takes a lot of beating. But the attack produced only one bomb crater in the runway at Port Stanley. The effort expended in making the attack was out of all proportion to the physical damage that it inflicted. Yet, as is often the case in aerial warfare, the psychological effect of the raid on the enemy was also out of all proportion to the physical damage that had been caused. The attack convinced the Argentine Air Force that Vulcans might strike at targets on the Argentine mainland at any time, and it withdrew its only specialized Mirage interceptor squadron to bases further north. The unit played little further part in the air fighting over the Falklands, and its withdrawal conceded air superiority in the skies over the islands to the Royal Navy Sea Harriers. That was the 'bottom-line' result of that first 'Black Buck' mission.

Reconnaissance is a vitally important aspect of air power and one that is sometimes neglected in accounts written on this subject. Yet the intelligence from a single photograph can have a major influence on the conduct of a land or an aerial bombing campaign. In Chapter 8, 'Reconnaissance over Normandy', we see how one Arado Ar 234 jet aircraft might have achieved decisive results had it gone into action just a couple of months earlier. In Chapter 15, 'Tornado Spyplanes Go to War' we see the state of the art of aerial reconnaissance as carried out during the recent war in the Persian Gulf. These aircraft carry no conventional film cameras. Instead, they use electro-optical, infra-red cameras similar to the camcorder to record the scene passing below the aircraft. The system is entirely passive and it works even on the darkest night. No longer is the main product of aerial reconnaissance the glossy photograph; now the buzz-word is 'electro-optical imagery'.

Chapter 14, 'Countdown to "Desert Storm"' brings together many of the lines of thought developed in previous chapters, to show the high-tech world of modern air warfare. The reader will see how the various disparate elements of the Coalition attack force came together during the first night's action in Operation 'Desert Storm'. The highly elaborate plan to neutralize parts of the enemy air defence system employed a spread of weapons ranging from armed helicopters, via cruise missiles, to F-117A stealth bombers. Scores of decoy drones were launched to bring the enemy missile batteries into action, then salvos of radiation-homing missiles were launched to knock out the missile guidance radars. In this chapter we also see the latest fighter escort techniques, in which powerful F-15 fighters were guided on to Iraqi fighters by E-3A AWACS planes orbiting far outside the battle area. That night teams of Royal Air Force Tornado fighter-bombers ran in at ultra-low altitudes to hit the runways and taxiways at Iraqi airfields with the purpose-built JP.233 airfield denial weapons. At the same time F-111s and A-6s attacked pin-point targets with the laser-guided and electro-optically guided bombs.

I have assembled this series of accounts in order to provide readers with as wide-ranging as possible an overview of the nature of aerial warfare. If after reading the accounts the reader feels that he or she has a better insight into the true nature of this complex subject, I shall consider that my efforts have succeeded.

Author's Note

Unless I have stated otherwise, throughout this account all miles are statute miles and all speeds are given in statue miles per hour. Gallons and tons are given in Imperial measurements. Where times are given, these are local for the area where the incident described took place. Weapon calibres are given in the units normal for the weapon being described, e.g. MK 108 30mm cannon or Browning .303in machine gun. Where an aircraft's offensive armament load is stated, this is the normal load carried by that type of aircraft during the operation described and not the larger maximum figure stated in makers' brochures and reproduced in most aircraft data books.

Alfred Price
Uppingham, Rutland

1

BY ZEPPELIN TO AFRICA

The huge airlift of troops and equipment to support the recent conflict in the Persian Gulf made unprecedented demands on Western military air transport fleets. But at least the crews knew where they expected to land before they took off, and they could be confident that they would not be called upon to fight as infantrymen when they reached their destination. It was quite different for those who set out for the first-ever intercontinental military airlift mission, almost three quarters of a century earlier . . .

IN SEPTEMBER 1917, the First World War was in its fourth year. In France and on the Eastern Front the opposing armies were locked in the deadly stalemate of trench warfare. Far away from all of this, in the German Protectorate of East Africa (now Tanzania), German and German-African troops were fighting a classic guerrilla-type war in the south-eastern highlands against overwhelmingly superior Allied forces. The German commander, *General* Paul von Lettow-Vorbeck, mounted occasional but sharp attacks then withdrew into the bush to avoid the larger forces sent in pursuit.

Thanks to the Allied naval blockade, the German troops in East Africa were desperately short of ammunition, weapons, medical supplies and every type of equipment. At the Colonial Office in Berlin one of the officials drafted an imaginative proposal for a method of circumventing the blockade and getting supplies through to the beleaguered force—by Zeppelin.

The idea was passed to the Imperial German Naval Airship Division for consideration, and when its staff officers investigated it they found that there were several major obstacles. First, the nearest German airship base to East Africa was at Jamboli (now Yambol) in Bulgaria, 3,600 miles away by the most direct route. No airship in existence could carry a useful load over so great a distance, though it might be possible to modify one of the craft under construction to do so. Second, since the Germans had no air base in Africa where the airship could refuel and replenish its supply of hydrogen, the flight would be a one-way mission. Third, this was a venture into unknown territory in more senses than one. No airship had ever attempted to fly so far over desert or tropical areas. It would encounter temperature variations between the hottest part of the day and the coldest part of the night that would be much greater than any previously experienced. These would induce large differences between the temperature of the hydrogen in the gas cells and that of the outside air, which could give rise handling problems not encountered before. Fourth, the maps of the interior of Africa were unreliable and accurate navigation would be difficult. And fifth, and not least, the Allies had several air bases in north Africa and in East Africa, which meant that the Zeppelin could come under attack from hostile aircraft.

Certainly the operation was a high-risk venture, but the risks were far outweighed by the physical and moral gains if the scheme succeeded. *Fregattenkapitän* (Captain) Peter Strasser, Chief of the Imperial Naval Airship Division, fervently endorsed the undertaking and commented:

> Completion of the operation will not only provide immediate assistance for the brave Protectorate troops, but will be an event which will once more enthuse the German people and arouse admiration throughout the world.

Admiral von Holtzendorff, Chief of Imperial Naval Staff, was equally enthusiastic and so was the German *Kaiser* when the proposals were submitted to him for final approval. Detailed planning for the operation, code-named 'China Matter', went ahead rapidly and under the strictest secrecy.

Had this been a normal operation, the Navy would have placed an experienced airship commander in charge of it. But if things went according to plan neither the airship nor any member of its crew would return in the foreseeable future, and when they reached their destination the crew of the Zeppelin would join the German ground forces in Africa and continue the fight as infantrymen. Always short of experienced commanders, the Naval Airship Service could not afford to lose one them in such a venture. As a result, a relatively junior and inexperienced Zeppelin commander, *Kapitänleutnant* (Lieutenant-Commander) Ludwig Bockholt, was chosen to take command of the 'China Matter'.

Zeppelin *L57*, whose construction was about to begin at Friedrichshafen, was chosen for the mission. To provide the extra lift necessary to carry the greater payload, the hull was lengthened by 99ft to make room for two additional gas cells. With a hydrogen capacity of 2,418,000 cu ft she was to be the largest airship yet built.

Several other changes were made to the Zeppelin to suit it for its unique mission. When it arrived in East Africa it was expected that the crew would deflate the craft and strip away everything that might be of use to the German forces. The radio transmitter and receiver, one of the engines, a dynamo and tanks containing the remaining petrol and lubricating oil were all to be removed and reassembled into a transportable ground radio station. Light alloy girders unbolted from the hull would support the aerials. Other girders would be removed and reassembled into a framework to support large areas of cotton fabric removed from the outer covering, to provide lightweight portable shelters for men and equipment. *L57* even had walkways around the hull made of leather—which could be unstitched and made into boots for the soldiers.

The much-modified airship made her maiden flight on 26 September 1917, but fate decreed that she would never leave Germany. Early the following month, following her fourth flight at the airship base at Jterbog near Berlin, a wind squall caught the Zeppelin and slammed her into the ground. Her back was broken and shortly afterwards the hydrogen exploded into flames. The crumpled structure sank to the ground and burned out. Fortunately for the crew, they had all scrambled clear of the airship before she caught fire.

Bockholt accepted the blame for the loss. He had been under pressure to complete the flight trials of the airship as quickly as possible, and in doing so he had ordered the flight to proceed despite an adverse weather forecast. A more experienced commander might have acted differently, but, as we have noted, a more experienced commander was not available. Moves to replace Bockholt were overruled at the highest level, and the young officer remained in command of 'China Matter'.

Meanwhile, in German East Africa events had also taken a turn for the worse. During September British troops cornered parts of von Lettow-Vorbeck's force and in the bitter fighting that followed both sides suffered serious losses. Radio communications between the German force and the homeland were poor, and the main source of information on the action available to Berlin came from intercepted enemy reports which spoke of an important British victory.

If 'China Matter' were to reap any of its sought-after benefits, it had to be undertaken quickly. Within two days of the loss of *L57*, the Naval Airship Service reassigned its next airship to be laid down, *L59*, to the Africa mission. Workers at the Staaken factory toiled round the clock to prepare the Zeppelin in a similar fashion to her sister-ship, and *L59* was completed in the remarkably short period of sixteen days. She made her maiden flight on 25 October, and successfully completed her trials. Then she was loaded with the 13¾ tons of supplies to be delivered to Africa, and she flew to her jumping-off base in Bulgaria on 4 November.

ZEPPELIN *L59*

Role: Transport airship; crew of 22.
Powerplant: Five Maybach HSLu petrol engines, each developing 240hp. One was mounted aft of the forward car, two were mounted in the rear car driving a single propeller via clutch mechanisms, and two were carried in the two separate engine cars suspended from the hull.
Armament: Several rifle-calibre machine guns were carried.
Performance: Maximum speed (all engines running) 64mph; cruising speed (one engine stopped for maintenance) 40mph.
Gas capacity: 2,418,700 cu ft held in 16 gas cells. This gave a gross lifting capability of 174,000lb under standard atmospheric conditions and a usable lifting capability of 114,400lb after the 59,600lb unladen weight of the airship structure and engines had been subtracted.
Dimensions: Length 743ft; diameter 78ft 5in.

Cargo carried by L59 during Africa operation (lb)

Small-arms ammunition, 381,000 rounds	21,777
20 machine guns and spare barrels	2,550
Madical supplies	5,790
Mail; miscellaneous items	770
Total	30,887

Consumables carried by L59 during Africa operation (lb)

Petrol	47,800
Lubricating oil	3,360
Water ballast	20,200
Total	71,360

Date L59 entered service: December 1917.

Still there was a dearth of information on the whereabouts of the main body von Lettow-Vorbeck's troops. Unless he received further instructions by radio, Bockholt's orders were to make for the area between Lake Nyasa and the Indian Ocean (close to the present-day border between Tanzania and Mozambique) and search for his countrymen there. If he found what looked like a body of friendly troops, he was to remain outside the range of small-arms fire while a petty officer volunteer jumped by parachute and made contact with the troops. If the man signalled confirmation that the troops were friendly, the airship was to descend.

After a couple of false starts, *L59* commenced her unique mission shortly after dawn on 21 November. At Jamboli the weather was almost perfect for the launch, with a northerly breeze of about 5mph and a ground temperature close to freezing point (the higher the temperature of the hydrogen in the cells compared with the ambient air, the greater the craft's lift). Each gas cell was filled to 95 per cent capacity, to allow room for the hydrogen to expand when the airship reached her cruising altitude. Fully loaded and ballasted, the airship was 'weighed off' with her lift and weight exactly balanced. On the order 'Airship out of the hangar—March!' the 400-strong ground handling party shuffled forwards, dragging the Zeppelin into the open like a gigantic broadsword being withdrawn from its scabbard.

Once the airship was clear of the shed the ground party halted, holding the craft stationary as her five engines stuttered into life. While they were warmed up, the big, 14ft diameter, two-blade propellers remained stationary in the horizontal position, their shafts declutched from the engines.

At 8.30 a.m., satisfied that all was ready, Bockholt barked the order 'Up ship'. At intervals along the length of the ship the command was repeated, and the ground party released its collective hold. At the pull of a toggle in the control car, 550lb of water cascaded from ballast bags to restore the Zeppelin to its lighter-than-air state. The craft drifted sluggishly upwards and once she was well clear of the ground the commander ordered the propellers to be engaged. There was a clanking of telegraphs in the engine cars, and the propellers started to turn. Slowly gaining speed and altitude, L59 edged round gently until she was pointing almost due south. The great adventure had begun.

For readers who are not familiar with the large rigid airship and its method of operation, a brief description of the craft might be appropriate at this point. From nose to tail the L59 measured 743ft—more than three times the length of a Boeing 747 jumbo jet—and her 91ft girth was more than four times that of the modern airliner. In addition to the 13¾ tons of supplies, the Zeppelin carried 21 tons of petrol, 1½ tons of lubricating oil and 9 tons of water ballast. With the crew of 22 living in spartan accommodation and supplied with food for ten days, in the fully loaded state the airship's gross weight was just over 77¼ tons.

Four gondolas hung below the hull of the airship, arranged in diamond fashion. The front gondola, or control car, housed the captain, the second-in-command, the radio operator, one rating to operate the rudders and another to operate the elevators. At the rear of the car was a hefty 6-cylinder Maybach HSLu petrol engine developing 240hp, driving a two-blade pusher propeller via a reduction gear. The two side gondolas each accommodated one of these engines driving a pusher propeller; and the rear gondola housed two such engines, which were clutched to the same shaft and drove a single pusher propeller. During normal operations each engine had a mechanic in constant attendance. One of the problems of the rigid airships was that they were seriously underpowered, and with a total of only 1,200hp the heavily laden L59 suffered from this affliction more than most.

CREW COMPOSITION OF L59

For the flight to Africa the airship carried a crew of 22:
Commanding Officer
Second-in-Command
Rudder operator: One rating plus relief.
Elevator operator: One rating plus relief. This crewman was also responsible for operating the gas valves and releasing ballast.
Navigator: one warrant officer.
Radio operator: One petty officer plus relief.
Senior Engineer: One warrant officer.
Engine room mechanics: One rating stationed at each of the five engines, plus five reliefs.
One sailmaker to make repairs on the gas cells and outer covering.
One medical doctor (not part of the normal crew of a Zeppelin).

The crew was divided into two watches. Those not on duty served as additional lookouts, or rested in hammocks strung across the main gangway that ran the length of the airship's hull.

Note: The big rigid airships were controlled in the manner of sea-going ships rather than aeroplanes. The officer of the watch stood on the bridge and made the flying control decisions, which were carried out by the two ratings who operated the craft's rudders and elevators. Control of the engines was ordered via telegraph indicators similar to those fitted to a ship, and carried out by mechanics in the engine cars.

Assisted by the moderate tail wind, the first twelve hours of the Zeppelin's flight were relatively uneventful. For two hours in every ten, each engine was shut down in turn for maintenance. Once the craft had left behind the mountainous area of eastern Turkey, the crew established her in the cruise at altitudes of about 2,000ft, maintaining an airspeed of about 40mph on the four running engines. A major difficulty facing Bockholt was the lack of any advanced meteorological information on conditions along the route. He was ignorant of the existence of difficult conditions until he ran into them, literally. Don Layton, an airship skipper who amassed some 4,000 flying hours aboard US Navy blimps between 1947 and 1957, described the problem:

It would have made all the difference if the crew could have received meteorological information by radio along the route. That would not have solved the problem of the head winds, but the crew could have mitigated their effect by re-routing or climbing or descending to avoid the worst of them. The crew could also have avoided areas where there were extremes of temperature, by routing around those places if they knew where they were. But they didn't have ground stations that could tell them if there was a hot spot ahead.

The lack of information soon made itself felt. On the first evening, as the airship headed over the Mediterranean, the weather suddenly deteriorated into an electrical storm. The wind rose, and to avoid the worst of the turbulence and the headwind Bockholt descended to 1,000ft. Rain and hail lashed the hull, then a sudden panic cry from the upper look-out station chilled the blood of all who heard it: 'The ship is on fire!' In fact it was only St Elmo's Fire, the shimmering, fluorescent blue glow caused by the build-up of static electricity. Large areas of the upper part of the hull were bathed in the strange light. It was the inexperienced lookout's first sight of the spectacle—which, despite its terrifying appearance, was quite harmless.

Shortly after dawn on the 22nd, the Zeppelin crossed the Egyptian coast near Sollum. The British possessed air bases in the Nile Delta, and to avoid these Bockholt kept well to the west. Once the ship was over the desert, identifiable ground features were few and far between, and the maps showing their positions were unreliable. The crew kept track of their movement southwards by a series of position lines from sun shots taken from the upper platform. The airship's drift and ground speed were calculated (with remarkable accuracy) by using a stopwatch to measure the time it took for its shadow to pass a point on the ground.

As the day progressed the crew encountered a set of problems quite different from those experienced when flying the airship over Europe. The hot sun playing on the hull heated the hydrogen, until that in some of the gas cells was 10°C warmer than the surrounding air. The phenomenon, known as 'superheating', caused the gas to expand and generate extra lift. In these circumstances the additional lift was an embarrassment, however. If the airship drifted above 2,500ft the pressure of the gas in the cells would exceed that outside the hull, and the automatic pressure release valves would then open and vent off some of the hydrogen. Like a miser hoarding gold, Bockholt had to minimize losses of the vital lifting gas. When the temperature fell, as it inevitably would that night, he would need all of his hydrogen.

To generate the negative dynamic lift to counter the unwanted lift, L59 flew nose-down over the desert. Don Layton described the problems of maintaining an airship in that flying attitude:

> L59 was 743 feet long, and if her nose went up or down a couple of degrees she produced a tremendous amount of drag. Then she would lose a lot of speed. If the nose of the craft was pushed down too far, it was easy to lose control. Something I used to enjoy doing with the pressure airships, if you are 'light' and flying nose down to prevent her gaining altitude and you are going too slowly, the airship would stall and drift upwards. It is the only vehicle I know where you can stall, and go up!

The manoeuvre that provided amusement for Layton during training flights would have cost Bockholt more hydrogen than he could afford.

Quite apart from that of maintaining the Zeppelin at the desired altitude, her crew had to contend with other problems. The hot air rising from the shimmering desert sands caused the craft to rise and fall gently but continually, like a cradle rocked by some giant invisible hand. The unfamiliar motion affected several crew members, and reduced even veteran seamen to the indignity of airsickness. In addition, the lookouts suffered blinding headaches from gazing for too long at the bright reflections off the sand.

During the afternoon the forward Maybach engine began to vibrate and had to be shut down. On opening the casing the mechanics found that the reduction gears had developed cracks that could not be repaired in flight. The motor could not be used again. With one 'good' engine always stopped for maintenance, the airship was to continue the flight on three engines. As luck would have it, the 'lost' engine drove the dynamo that supplied power to the airship's radio transmitter, which meant that the latter could not be used either.

Still the airship continued determinedly southwards, and late that night she passed over the Nile at Wadi Halfa. By midnight L59 was well into the Sudan and approaching the latitude of Khartoum. Since leaving Jamboli she had covered 2,800 miles and was more than half way to her destination. Despite the baking heat of the day, followed by sub-zero temperatures at night, the crew made light of their physical discomfort. The prospects of success increased with each hour that passed and they knew that their triumphant arrival would give new heart to the Fatherland and its allies, and cause dismay to their foes.

Yet circumstances now conspired to defeat the best efforts of the doughty crew. Intercepted British radio reports reaching Berlin painted a disturbing picture of Allied troops thrusting into parts of German East Africa that had previously been considered secure. This, coupled with the lack of information on the location of von Lettow-Vorbeck's forces, gave little chance of the Zeppelin making a successful rendezvous. Reluctantly, the Imperial Naval Staff gave orders for the recall signal to be broadcast to L59:

Break off the operation, return. The enemy has captured the greater part of the Makonde Highlands, also Kitangari. The Portuguese are attacking the Protectorate from the south.

The Zeppelin's radio operator picked up the message but, lacking power for his transmitter, he could not acknowledge its receipt. The plaintive signal would be repeated at regular intervals during the days to follow, while in Berlin there were mounting fears for the safety of *L59* and her crew.

Aboard the Zeppelin, the recall signal caused shock and dismay. The crew had risked so much and achieved so much, for nothing. No longer buoyed by the adrenalin of impending glory, several men succumbed to the discomforts and developed symptoms of nervous tension and feverishness. Worse followed, and soon. As the airship headed back for Europe, a problem that had been building up since nightfall neared crisis proportions. After sunset the temperature of the hydrogen fell rapidly, giving rise to 'supercooling', the reverse what had happened during the day. As the hydrogen slowly cooled, its volume contracted, and gradually the Zeppelin ceased to be a 'lighter-than-air' craft.

To maintain the 3,000ft altitude necessary to clear high ground in the area, Bockholt ordered the helmsman to hold the craft in a 4-degree nose-up attitude, with all four usable engines running at full power. The release of 4,400lb of water ballast eased the problem of maintaining altitude, but not for long. The process of 'supercooling' continued, until in some cells the hydrogen was 6°C lower than the surrounding air. Never before had an airship encountered so great a swing in temperature in such a short time.

Matters to a head shortly before 3 a.m. that morning. Without the crew realizing it, the airship ran into a layer of relatively warm air, which further decreased the lift from the hydrogen. When the helmsman increased the nose-up attitude to hold the required 3,000ft altitude, the Zeppelin suddenly stalled and began to fall out of control. She came down relatively slowly, at about 150ft a minute, but the situation was no less serious for that. Don Layton outlined the problem:

> An airship has a lot of area when viewed from below. When it was coming down under those conditions it was like a huge inflated 'parachute'. It did everything slowly, but you had no control over it. Once you'd lost the airflow over the fins you didn't have any rudder control and elevator control was sloppy. Although the airship was going down relatively slowly, unless something was done she wasn't going to stop. Once she started going down, she had an awful lot of momentum.

The knee-jerk reaction would have been a panic release of ballast to lighten the airship, but her commander had to be much more circumspect. Layton described a flight in a free balloon as part of his training as an airship pilot, when he went down out of control:

> You start coming down so you throw out ballast. You still go down so you throw out more. And still you go down. I threw out the rest of the ballast,

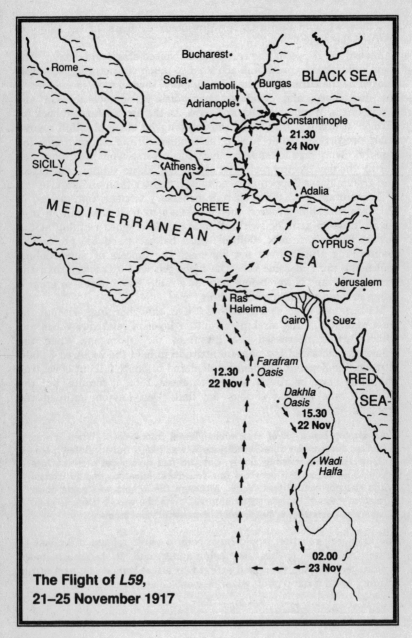

The Flight of *L59*,
21–25 November 1917

then the instrument panel, then our flight jackets. I thought it was never going to stop going down until it hit the ground. Finally it did stop. But by then we had thrown out too much, and the balloon started going up again! That was what happened if you got behind what the craft was doing. It would have been even worse for a big craft like the *L59*, particularly if she was coming down nose first or tail first.

Bockholt took the appropriate stall recovery action for an airship, ordering the release of some ballast and having the propellers declutched from the engines until he had recovered control. When these measures failed to arrest the descent, he jettisoned part of the cargo. Only after the airship had been lightened by 6,600lb, having narrowly missed smashing into the side of a hill on the way, was its downward drift halted. When *L59* bottomed out she was uncomfortably close to the ground, in a ravine with steep-sided and craggy ridges on either side. As she began to float upwards, her commander ordered the propellers to be engaged and as the craft slowly picked up speed he gingerly manoeuvred her away from the danger. It has been a mouth-drying experience for all on board.

L59 retraced her earlier path across Egypt and left the coast early in the afternoon of the 24th. She crossed the Mediterranean without incident, and darkness had fallen when she began to re-cross the mountains of eastern Turkey. And there the crew had another brush with death. The Zeppelin suffered a re-run of the near-catastrophe over the Sudan, and for the much same reason: flying 'heavy' at night, relying on dynamic lift to maintain altitude, she caught a downdraught and started to descend out of control. Only after the release of 3 tons of ballast and ammunition was control restored.

The airship arrived at Jamboli at 4.30 a.m. in the morning of the 25th. Nobody there had anticipated that she would return and, since her radio transmitter out of action, she appeared unannounced. *L59* had to circle the base for a couple of hours until the necessary landing party was assembled, and not until 7.40 a.m. was she finally settled into the hands of the ground crew. Mentally and physically exhausted, the captain and crew climbed slowly from the craft that had brought them back safely but unexpectedly.

The attempt to transport supplies to the German troops in East Africa had failed, but that was no fault of *L59* or her crew. The men had done all that had been asked of them, and they could indeed be proud of their achievement: in a flight lasting 95 hours 5 minutes, just under four days, they had covered a distance of 4,200 miles—about as far as from London to Miami. That distance would have easily taken them to any rendezvous with von Lettow-Vorbeck, if he could have been found. The flight had broken every distance record by a wide margin, with a cargo heavier than any previously transported by air. And the crew could have gone a long way further had they been required to do so: the 2,760 Imp gallons of petrol remaining in the Zeppelin's tanks when she landed at Jamboli could have kept her engines running for a further 2½ days.

In time of peace Ludwig Bockholt and his men would have been fêted as national heroes, but this was wartime and other long-distance missions were being considered for the modified Zeppelin. Her flight remained shrouded in secrecy and would not be revealed until after the war. In East Africa the German commander succeeded in rallying his forces after the hard fought-battles during the autumn, and they continued to mount occasional attacks until the end of the war. There was no attempt to resurrect the idea of supplying the troops by Zeppelin, however.

Neither *L59* nor most of her intrepid crew would long outlive their return from Africa. The airship reverted to normal bombing operations and on 7 April 1918 she set out from Jamboli for an attack on the Royal Navy base at Valletta, Malta. On the way to the target the Zeppelin suddenly burst into flames and plunged into the sea. There were no survivors. No Allied forces in the area reported any engagement that can be linked to the incident. A subsequent German investigation concluded that '*L59* was probably lost to accidental causes, possibly as a result of leaking petrol catching fire and igniting the hydrogen'.

In summing up the flight of *L59* to Africa and the performance of her crew, Don Layton commented:

I think the flight was one of the great epics of aviation. In terms of hours flown it was the longest operational flight, ever. That would have imposed considerable stress on the crew over a very long period. On the way out everybody was buoyed up, full of adrenalin and ready to go. But I think that on the return flight their morale must have been very low. Another point to consider is that Bockholt was an inexperienced airship commander, and his second-in-command was even less experienced. It is likely that the captain had to stay close to the controls throughout almost the entire four-day flight. I take my hat off to Bockholt; I think they ought to put up a monument to him some place.

2

BATTLE OF BRITAIN DAY: 15 SEPTEMBER 1940

In August 1940 the Royal Air Force attacked targets in Germany, in retaliation for bombs that had fallen on London. The move enraged Hitler, who ordered a massive bombardment to be unleashed against London by way of reprisal. The new phase in the Battle of Britain opened on 7 September, when several hundred bombers attacked the dock areas to the east of the city, causing severe damage and several large fires. In the week that followed there were three further daylight attacks on the capital. For the fifth action in the series, to take place on Sunday 15 September, the Luftwaffe *planned to deliver two separate attacks on the city, both with a strong escort of fighters. The resultant action would mark the climax of the Battle of Britain.*

DURING THE FIRST four German daylight attacks on London, on each occasion a large proportion of the defending fighters had failed to come to grips with the raiding formations. The initial attack had surprised the defenders and the fighters had been positioned to meet yet another attack on their airfields. During the next three daylight attacks on the capital, banks of cloud hindered the tracking of aircraft over southern England by ground observers; the result was a series of scrappy actions with relatively light losses on both sides.

Luftwaffe intelligence officers ascribed a quite different reason to the lack of a ferocious fighter reaction that characterized several of the actions in August: they thought it was a symptom of the long-predicted collapse of RAF Fighter Command. If the latter assessment were accurate, the correct German strategy was to mount further large-scale attacks on London that would force the surviving British fighters into battle to suffer further losses from the escorting Messerschmitts.

Seeking to follow up its apparent advantage, the *Luftwaffe* planned to mount two separate attacks on the British capital for 15 September, one against an important part of the rail system and the other against the docks. Every Messerschmitt Bf 109 unit in *Luftflotte 2* was to be employed in supporting these attacks, several flying double sorties.

In fact, in the second week in September Fighter Command was numerically as strong as it had been when the Battle began. German intelligence officers had underestimated British fighter production, and also the ability of the repair organization to get damaged fighters back into action. Moreover, although Fighter Command had lost several of its experienced pilots, there had been a steady infusion of these from other commands and also from the air forces of countries under German occupation. The force had also gained considerable combat experience in the course of the Battle. The action about to take place would be no 'walk-over' for the *Luftwaffe*.

The first attack on the capital was to take place during the late morning, with a raiding force comprising twenty-one Messerschmitt Bf 109 fighter-bombers of *Lehrgeschwader 2* and twenty-seven Dornier 17s of *Kampfgeschwader 76*. Providing open and close-cover escort for the bombers were about 180 Bf 109s. The fighter-bombers were to hit rail targets, then the Dorniers were to deliver a precision bombing attack on the important swathe of railway lines and viaducts running through the borough of Battersea.

Soon after the first of the German bombers got airborne, however, this part of the day's operation started to go awry. As the Dorniers climbed away from their base airfields heading for the Pas de Calais to link up with the fighter escort, the aircraft ran into a layer of cloud that was considerably thicker than forecast. *Feldwebel* Theodor Rehm, a navigator in one of the bombers, described what happened next:

> In cloud the visibility was so bad one could see only the flight leader's plane a few metres away. In our bomber four pairs of eyes strained to keep the aircraft in sight as its ghostly shape disappeared and re-appeared in the alternating darkness and light. One moment it was clearly visible, menacingly large and near; then suddenly it would disappear from view, in the same place but shrouded in billowing vapour. At the same time we also scanned the sky around for other aircraft, ready to bellow a warning if there were a risk of collision. After several anxious minutes that felt like an eternity, we emerged from cloud at about 3,500 metres [11,000ft]. Around us were the familiar shapes of Dorniers spread over a large area in ones, twos and threes. Our attack formation had been shattered.

Major Alois Lindmayr, the German formation leader, flew a series of wide orbits to allow the others to re-form behind him. Two of the bombers failed to rejoin the formation and returned to base.

Ten minutes late, the remaining bombers reached the Pas de Calais and picked up their escorts, then swung on to a north-westerly heading for London. The Messerschmitts of the close-cover and open-cover forces held position around the bombers, while those assigned to the free-hunting patrols sped out in front of the force. Now for a second time the weather intervened in the proceedings to the raiders' disadvantage, for at the bombers' altitude of 16,000ft there was a fierce 90mph headwind.

Running in at altitudes around 23,000ft, the Bf 109 fighter-bombers reached London unchallenged by the defenders. As intended, the high-flying Messerschmitts looked like any other German free-hunting patrol, and, given the choice, RAF pilots preferred to leave those well alone. Only one of the defending pilots' reports, that from Pilot Officer P. Gunning of No 46 Squadron, can be linked to this incursion. He reported seeing a force of Messerschmitt 109s pass over him, but these aircraft '. . . did not appear to attempt to attack anyone below.' Certainly the German fighter-bomber pilots had no intention of becoming embroiled with enemy fighters unless they were directly threatened. It was one of those confrontations in which both sides were quite happy to leave the other unmolested.

The fighter-bombers reached the city and their leader, *Hauptmann* Otto Weiss, banked his aircraft steeply and picked out one of the smaller urban railway stations. Then, with his two wing-men maintaining close formation on either side, he pushed his Messerschmitt into a 45-degree dive and lined up on the target. After a descent of just over 3,000ft, Weiss released his 550lb bomb and his wing-men released theirs, then all three planes pulled out of their dives. The other Messerschmitts in the force delivered their attacks in a similar manner. It was an inaccurate method of aiming, however, and most of the bombs fell across residential areas in the boroughs of Lambeth, Streatham, Dulwich and Penge. The raiders caused little damage and few casualties, then withdrew without loss.

While this was happening, the Dorniers and their escorting Messerschmitts fought their way across Kent to get to the capital. Throughout this time the British fighter controllers fed eleven squadrons of Spitfires and Hurricanes into the action in ones and twos, to set up a series of skirmishing actions around the raiding force. The

---

SUPERMARINE SPITFIRE Mk I

Role: Single-seat fighter.
Powerplant: One Rolls-Royce Merlin III 12-cylinder, liquid-cooled engine developing 1,030hp at 16,250ft.
Armament: Eight Browning .303in machine guns mounted in the wings.
Performance: Maximum speed 353mph at 20,000ft; climb to 20,000ft, 7min 42sec.
Normal operational take-off weight: 6,050lb.
Dimensions: Span 36ft 10in; length 29ft 11in; wing area 242 sq ft.
Date of first production Spitfire I: May 1938.

Messerschmitts responded energetically to each attempt to break through to their charges, however, and the bombers reached the eastern outskirts of London without suffering a single loss.

Now, however, the delay the Dorniers had incurred when they reassembled their formation over France, coupled with the powerful headwind experienced over Kent, meant that the bombers arrived in the target area more than half an hour late. The twin-engine aircraft carried sufficient fuel to cope with such an eventuality, but the Messerschmitts did not. For the latter, even under optimum conditions, the British capital lay close to the limit of their effective operational radius of action when flying from bases in France. Now they were running low on fuel, and the escorts were forced to break away from their charges and turn for home. As the Dorniers lined up for their bombing run on Battersea, the last of the escorts departed.

In the No 11 Group command bunker at Uxbridge, Air Vice-Marshal Keith Park had no way of knowing his enemy's predicament. He had already decided to fight his main action over the eastern outskirts of the capital, however, and had directed seven squadrons of Spitfires and Hurricanes to that area to engage the raiding force. In addition, from No 12 Group to the north, Squadron Leader Douglas Bader was making for the capital at the head of a 'Big Wing' with five further squadrons of fighters.

The main action began shortly after mid-day, 16,000ft above Brixton. Holding his Dornier straight and level for the bomb run, *Feldwebel* Wilhelm Raab noticed what looked like a swarm of small flies emerge from behind cloud ahead of him:

> Of course they weren't flies. It was yet more British fighters, far in the distance but closing rapidly. I counted ten before I had to give up and concentrate on holding formation.

The pilot of the nearest 'small fly', Squadron Leader John Sample, was leading twenty Hurricanes of Nos 504 and 257 Squadrons in for the attack. Ahead of him he could see the Dorniers clearly, silhouetted against the tops of the clouds:

> As we converged I saw that there were about twenty of them and it looked as though it was going to be a nice party, for the other squadrons of Hurricanes and Spitfires also turned to join in. By the time we reached a position near the bombers we were over London. We had gained a little height on them, too, so when I gave the order to attack we were able to dive on them from their right.

To Raab and many of those in the Dorniers, it looked as if the enemy fighter pilots had waited until the Messerschmitts turned away before

DORNIER Do 17Z

Role: Four-seat bomber.
Powerplant: Two Bramo 323 Fafnir 9-cylinder, air-cooled radial engines each developing 1,000hp at take-off.
Armament: Normal operational bomb load 2,200lb. Defensive armament of six Rhienmetall Borsig MG 15 7.9mm machine guns in flexible mountings, two firing forwards, two firing rearwards and one firing from each side of the cabin.
Performance: Maximum speed 255mph at 13,120ft; normal formation cruising speed at 16,000ft, 180mph; radius of action with normal bomb load, 205 miles.
Normal operational take-off weight: 18,930lb.
Dimensions: Span 59ft 0¼in; length 51ft 9½in; wing area 592 sq ft.
Date of first production Do 17Z: Early 1939.

they delivered their main attack. The truth was more prosaic: Sample and the other British squadron commanders were ignorant of the Dornier crews' problems and were merely following the intercept instructions from their ground controllers. The fact that there were no escorts in the area was a lucky break for the RAF pilots, but from better experience they knew that the Messerschmitts had a nasty habit of turning up at the most inconvenient time. Everyone maintained a careful lookout.

Despite the sudden disappearance of the fighters that should have been escorting his Dorniers, Alois Lindmayr held his heading for the target. John Sample and his pilots picked out individual bombers in the formation and closed on them swiftly, firing long bursts. As they broke away, the Hurricanes of No 257 Squadron delivered their attack. Then the fighters split up into twos and threes and curved tightly to get behind the bombers for further attacks.

The Dornier crews held tight formation and traded blows with their tormentors, until the inevitable happened: one of the bombers suffered engine damage and was forced to drop behind the formation. The straggler immediately came under attack from several fighters and was badly shot up. Three of the crew, including the pilot, bailed out.

Things were going badly for *Kampfgeschwader 76* and they were about to get worse, for now Douglas Bader's 'Big Wing' arrived over the capital with its fifty-five fighters. The size of the force would later swell with the telling, and the official *Luftwaffe* report on the action noted: 'Over the target large formations of fighters (with up to 80 aircraft) intercepted.' For bomber crews who had been confidently assured a few hours earlier that Fighter Command was in its death throes, it was a devastating sight.

As Bader closed on the enemy force, the Dorniers were in the final seconds of their bombing run. Wilhelm Raab observed:

With the British fighters whizzing through our formation, the leading aircraft began releasing their bombs. My navigator shouted '*Ziel!*' and released ours.

Each bomber in the formation disgorged a stick of twenty 110-pounders, then, lighter by about a ton, it began a sweeping curve to the left.

Flying on autopilot with two dead or dying crewmen on board, the Dornier that had left the formation continued on its north-westerly course across the capital. Four Hurricanes and a Spitfire from three different squadrons engaged the bomber, then John Sample ran in to deliver a further attack. Later he wrote:

> I found myself below another Dornier which had white smoke coming from it. It was being attacked by two Hurricanes and a Spitfire, and was travelling north and turning slightly to the right. As I could not see anything else to attack at that moment I climbed above him and did a diving attack. Coming in to the attack, I noticed what appeared to be a red light shining in the rear gunner's cockpit, but when I got closer I realised I was looking right through the gunner's cockpit into the pilot's and observer's cockpit beyond. The 'red light' was a fire. I gave it a quick burst and as I passed him on the right I looked in through the big glass nose of the Dornier. It was like a furnace inside.

Finally the lone Dornier was finished off in unconventional fashion by another member of Sample's squadron. Sergeant Ray Holmes ran in to deliver a head-on attack, but after a short burst his Hurricane's guns fell silent—he was out of ammunition. The fighter pilot made the snap decision to continue on and ram the enemy plane, and a split second later his port wing struck the rear fuselage of the Dornier. The collision sliced off the bomber's entire tail and, deprived of the balancing force provided by that vital appendage, the plane's nose dropped violently. The forward momentum maintained the aircraft on its previous flight path, however, imposing tremendous forces on the wing and causing the structure to fail. Both outer wing sections sheared away just outboard of the engines and the doomed bomber entered a violent spin.

The incident took place three miles high, almost exactly over Buckingham Palace. The Dornier still carried its bomb load, and the savage *g* forces now imposed further unsustainable stresses on the plane's already weakened structure. Two 110lb bombs and a container of incendiary bombs tore away from their mountings, smashed their way out of the aircraft and fell into space. One 110-pounder plunged into the roof of the Palace and passed through a couple of floors before

By Zeppelin to Africa

Aʙᴏᴠᴇ: The 22-man crew of *L59*, pictured after their return from the epic flight to Africa. *Kapitänleutnant* Ludwig Bockholt, the Captain of the airship, is in the centre of the rear row wearing the dark hat and dark leather coat. (IWM)

Battle of Britain Day: 15 September 1940

Bᴇʟᴏᴡ: Taken during the morning of 15 September, this photograph depicts a Dornier Do 17 crew of *Kampfgeschwader 76* at Beauvais before taking off to attack London. Just over two hours later the bomber was shot down by British fighters. *Oberleutnant* Wilke and *Feldwebel* Zrenner (second from left and on right, respectively) bailed out and were taken prisoner; the others were killed. (Rehm)

BELOW LEFT: Its tail knocked off when it was rammed by Ray Holmes, this Dornier did a sudden bunt and both outer wing panels broke away. Then the bomber fell out the sky in a vicious spin. Ray Holmes's Hurricane is seen below it, diving towards the ground out of control.

BELOW RIGHT: The severed tail of the Dornier falling out of the sky over central London. It came down on the roof of a house in Horseferry Road.

Above: Sergeant Ray Holmes of No 501 Squadron, who was fortunate to escape with his life. (Holmes)

Right: *Oberleutnant* Peter Schierning of *Kampf-geschwader 53*, whose Heinkel He 111 was damaged by flak near Chatham and forced to leave its formation during the afternoon action on 15 September. The bomber was finished off by Hurricanes. (Schierning)

Pre-Emptive Strike

LEFT UPPER: *Leutnant* Diether Lukesch of *III Gruppe* of *Kampfgeschwader 76*, left, whose description of the attack on Krudziai airfield in Lithuania on 22 June 1941 is included in this account. He and his observer, *Leutnant* Dipesen, are pictured in front of their Junkers Ju 88 bomber. (Lukesch)

LEFT LOWER: A view of Krudziai airfield as the Ju 88s of KG 76 ran in to deliver their attack. Although it took place several hours after the start of the German offensive, the attack achieved complete surprise. Note the Tupolev SB-2 bombers still drawn up in a neat line in front of the woods. (Lukesch)

ABOVE: Schaulen airfield in Lithuania, pictured after its capture by German troops. In the foreground is a Polikarpov I-16, the most numerous Soviet fighter type in the summer of 1941. The damage to the wing of the SB-2 bomber on the far right of the photograph is consistent with having been caused by a direct hit by an SD-2 fragmentation bomb. Visible above the tail of the I-16 is a Gloster Gladiator, one of several that were supplied to the Lithuanian Air Force before the war. (Lukesch)

BELOW LEFT: The SD-2, a 4½lb fragmentation bomb used in large numbers during the campaign against the Soviet Union. These weapons sometimes proved dangerous to their users, and in the *Luftwaffe* they were nicknamed 'Devil's Eggs'.

BELOW RIGHT: A low-flying German aircraft releases a line of SD-2 bombs across a Soviet airfield. In the foreground is a Polikarpov I-15 biplane fighter and a refuelling vehicle. (Dierich)

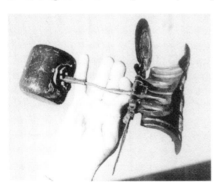

Guided Missiles Make Their Mark

Previous spread: A Royal Air Force Liberator patrol aircraft, one of the types that carried the 'Mark 24 Mine' homing torpedo into action. To force U-boats to submerge so that they can be attacked with the new weapon, this aircraft carries launchers for eight 60lb rockets on the sides of the fuselage.

Above: A Dornier Do 217 of *II Gruppe* of *Kampfgeschwader 100*, with a Henschel Hs 293 glider bomb fitted to the rack under the starboard outer wing. (Via Girbig)

Below: A dramatic still from a cine film taken by a Royal Navy officer during a glider-bomb attack on his ship. The missile missed—but not by much!

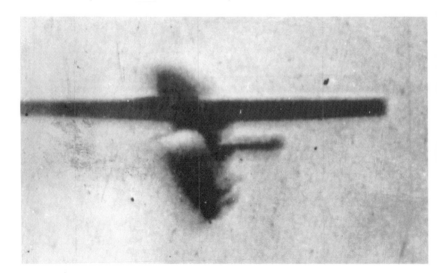

Above: The Royal Navy frigate *Egret* exploding on 27 August 1943, after a hit by a glider bomb set off the ammunition in her after magazine. (Selinger)

Below: A Fritz-X guided bomb pictured under the fuselage of a Heinkel He 177 bomber. Released from an altitude of about 20,000ft, this unpowered weapon would achieve an impact velocity close to the speed of sound.

Bildstelle
III. K.G. 100.

C = Korper
---- = Leuchtsatzschweif
O = Einschlag

Volltreffer auf Ital. Schlachtschiff
Klasse „Littorio" 35000 t
Aufgen. u. geworfen am 9.9.43 + 15.40
Beob. Uffz. Degan
Maßstab etwa 1: 9400

C.

Treffer
auf Achterdeck

Left: A photograph taken in the afternoon of 9 September 1943, during the attack by Dornier Do 217s on the Italian battleship *Roma*. Note that the ship is evading in a high-speed turn to port. One bomb has already hit her on the afterdeck, and another is being guided towards her from the right (the tracking flare is marked with a dotted line and a semi-circle). (Jope)

Hard Fight to 'The Big B'

Below: Colonel 'Hub' Zemke led the escorting P-47 Thunderbolts of the 56th Fighter Group during the action on 6 March 1944. During the initial combat over Haselünne he shot down the FW 190 flown by *Leutnant* Wolfgang Kretchmer of *II Gruppe, Jagdgeschwader 1*. (USAF)

Left: Kretchmer bailed out of his blazing fighter and parachuted to safety. (Kretchmer)

Below: 'Flak you could walk on', seen from a B-24 Liberator of the 453rd Bomb Group over Berlin on 6 March 1944. (Cripe)

Right upper: A stick of twelve 500-pounders falling from a Liberator of the 446th Bomb Group. At the top of the picture is Lake Harvel, one of several stretches of water around Berlin. (USAF)

Right lower: A deceptively peaceful picture of B-17 'Flakstop' of the 452nd Bomb Group as she flies past Bremen on her way home from Berlin on 6 March 1944. About ten minutes after this photograph was taken the bomber was shot down by German fighters. (USAF)

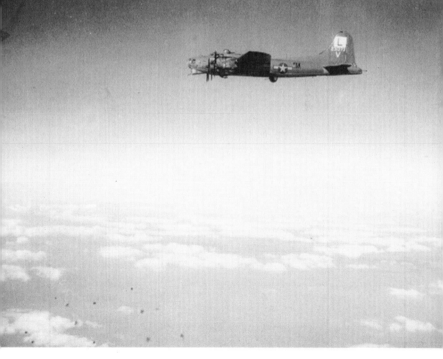

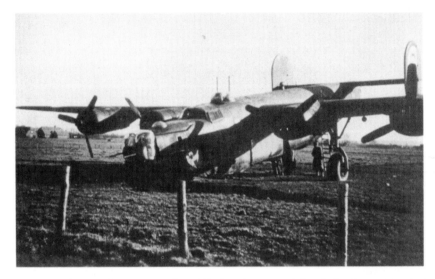

Above: B-24 Liberator 'Flak Ducker' of the 392nd Bomb Group, pictured after she crash-landed near Amersfoort in Holland. All her crew were taken prisoner.

The Great Marianas Turkey Shoot
Below: The Mitsubishi A6M5 Model 52 ('Zeke 52'), the standard Japanese Navy carrier fighter in the summer of 1944, was inferior to the Hellcat in almost every aspect of performance that mattered in fighter-versus-fighter combat.

Reconnaissance over Normandy
Right: *Leutnant* Erich Sommer flew the world's first jet reconnaissance mission on 2 August 1944 when he piloted an Arado Ar 234 to photograph almost the entire Allied lodgement area in Normandy. (Sommer)

Above: An Ar 234 prototype pictured shortly after lifting off from its unusual take-off trolley. The parachute to slow the trolley has already started to deploy. The lack of such a trolley at Juvincourt delayed the first jet reconnaissance mission by one week.

Below: One of the photographs taken during Eric Sommer's reconnaissance flight over Normandy on 2 August 1944, showing the Allied 'Mulberry Harbour' off Asnelles-sur-Mer.

coming to rest in the bathroom of one of the Royal Apartments. The other 110-pounder, and the container with sixteen incendiary bombs, landed in the Palace grounds. The high-explosive bombs failed to detonate but some of the incendiary bombs ignited on impact. The main part of the Dornier came down on the forecourt of Victoria Station, while the severed tail unit fell on a house in Vauxhall Bridge Road nearby. Borne on the strong north-westerly wind, the outer wing sections fluttered down slowly and covered more than a mile before they landed south of the Thames in Vauxhall. Ray Holmes's Hurricane did not emerge from the collision unscathed either. The port wing suffered severe damage and the fighter went out of control, and with some difficulty the pilot struggled out of his cockpit. The fighter plummeted into a road junction in Chelsea and Holmes landed by parachute in Pimlico.

Remarkably, considering that the bombs and the wreckage of both aircraft came down on built-up areas, nobody was injured on the ground and there was relatively little damage to property. The incendiary bombs starting small grass fires in the grounds of Buckingham Palace, but these were quickly extinguished by Palace staff. No member of the Royal Family was in the Palace at the time. Two of the crewmen who bailed out of the Dornier were taken into captivity soon after landing. The third man, the pilot, was less fortunate. He landed by parachute beside The Oval Underground station in Kennington, but before troops could reach him he was attacked by angry civilians and suffered fatal injuries.

As the formation of Dorniers withdrew from the capital it was under sustained attack from more than a dozen squadrons of fighters. Nine bombers suffered damage and were forced out of formation, and five of them were finished off shortly afterwards by fighters. Near Maidstone the Bf 109s assigned to cover the bombers' withdrawal linked up with them and shepherded the survivors home, assisted by the same 90mph wind that had impeded the raiding force on its way in.

Kampfgeschwader 76 had suffered a fearful mauling. Of the twenty-five bombers that had crossed the coast of England less than 50 minutes earlier, six had been shot down and four had been seriously damaged and only just made it back to France. Most of the fifteen Dorniers that stayed in formation had also taken lesser amounts of battle damage. Yet, considering the absence of escorts over the target and the overwhelming concentration of RAF fighters present, it is perhaps surprising that any of the Dorniers survived. The fact that three-quarters of the bombers managed to return to France is testimony to the tactical leadership of Alois Lindmayr and the flying skill and discipline of his crews. It is also testimony to the ruggedness of the Dornier 17's structure and its ability to withstand heavy punishment.

In contrast to Fighter Command's reaction to previous attacks on the capital, the system of fighter control that morning had functioned brilliantly: twenty-three squadrons with a total of 254 Spitfires and Hurricanes had been scrambled, and every squadron except one had made contact with the enemy.

As the survivors of the first attack on London crossed the coast of France, the bombers assigned to the second attack on the capital were airborne and heading for the Pas de Calais to rendezvous with their escorts. Far larger than the previous raiding force, this one comprised 114 Dorniers and Heinkel 111s. Their targets were in the Port of London to the east of the city—the Royal Victoria, the West India and the Surrey Commercial Docks. About 360 German fighters were to support the incursion.

As soon as the incoming German force appeared on his plotting table, Air Vice-Marshal Keith Park initiated his riposte. Eight squadrons of Spitfires and Hurricanes were ordered to scramble and patrol in pairs over Sheerness, Chelmsford, Hornchurch and Kenley. During the ensuing minutes other squadrons followed them into the air.

After crossing the coast at Dungeness, the attack force wheeled on to a north-north-westerly heading, making for the capital. The raiders were arrayed in five separate *Gruppe*-sized formations, the smallest with eighteen and the largest with 29 bombers. The three leading formations flew in line abreast, three miles apart; the other two flew three miles behind the left and the right leading formations.

Soon after the bombers crossed the coast, three of the forward-deployed Spitfire units, Nos 41, 92 and 222 Squadrons, with a total of twenty-seven fighters, went into action. The escorting Messerschmitts moved into position to block the attacks and one of the German pilots, *Hauptmann* Fritz Losigkeit of *JG 26*, later recalled:

After we crossed the coast the British fighters came in from a great height, going very fast. They broke through to the He 111s ahead of us and below, to attack the rear of the formation. During the dive some of the Spitfires became detached from the others. Using full throttle, my *Staffel* was able to catch up with them and I get into an attacking position. I fired a long burst and pieces broke away from the Spitfire's wing and fuselage. The pilot slid back the canopy and jumped from the cockpit. Overtaking rapidly, I pulled to the left of the Spitfire and saw his parachute open.

Probably the Spitfire that Losigkeit hit was that flown by Pilot Officer Bob Holland of No 92 Squadron. The latter suffered minor injuries on landing.

As the first of his fighters went into action, the last of Keith Park's day fighter units were taking off. No 11 Group had twenty-one squadrons of Spitfires and Hurricanes and every one was now airborne. From No 12 Group in the north, Squadron Leader Douglas Bader was again on his way south at the head of the five-squadron 'Big Wing'. From No 10 Group in the west, three squadrons were on their way to meet the raiders. Fighter Command had now committed all 276 of the Spitfires and Hurricanes based close enough to the capital to assist with its defence. Winston Churchill was at Park's underground operations bunker at Uxbridge that day and he watched as the No 11 Group commander committed his entire force. Later the Prime Minister wrote:

I became conscious of the anxiety of the Commander, who now stood still behind his subordinate's chair. Hitherto I had watched in silence. I now asked: 'What other reserves have we?' 'There are none,' said Air Vice-Marshal Park. In an account which he wrote about it afterwards he said that at this I 'looked grave'. Well I might. What losses should we not suffer if our refuelling planes were caught on the ground by further raids of '40 plus' or '50 plus'! The odds were great; our margins small; the stakes infinite.

In this action the British fighters were outnumbered by more than two to one by the raiders; in terms of single-seat fighters, there were three Bf 109s airborne over southern England for every two Spitfires and Hurricanes. Keith Park's tactics during the afternoon engagement were the same as those he had employed in the morning. As the raiders tracked across Kent he again fed fighter squadrons into action in pairs. In one of these encounters Nos 213 and 607 Squadrons, with twenty-three Hurricanes, delivered a head-on attack on the Dorniers of *Kampfgeschwader 3*. During this action there was another case of deliberate ramming, as Pilot Officer Paddy Stephenson of No 607 Squadron recalled:

Our squadron was flying in four vics, in stepped-down formation. The bombers were flying in stepped-up formation. In a head-on attack each vic was supposed to pass above the aircraft being attacked, and immediately below the following bomber. To do this there needed to be a proper spacing between the vics in our squadron. My vic had moved too far forward and to break away upwards would have involved me in a crash with a Hurricane in the leading vic of our formation.

Stephenson made a quick decision: if there had to be a collision, it was far better to hit a foe rather than a friend. He held his heading towards the bomber and the Hurricane's starboard wing smashed into the starboard wing of one of the Dorniers, shattering both. *Feldwebel* Horst Schultz, one of the German bomber pilots, watched the approaching Hurricanes with wings blinking as they opened fire.

> The next moment there was an explosion in front of me, then pieces of planes were falling out of the sky like confetti. I didn't know whether a British fighter had collided with one of our planes, or if it had suffered a direct hit from flak. There was no time to brood about it: I had my own job to do.

The rammed Dornier immediately entered a vicious spin, the savage *g* forces pinning the terrified crew in the cabin until the bomber plunged into a wood near Kilndown. Deprived of a large section of one wing, Stephenson's Hurricane rolled on to its back, then went down in a steep inverted dive. After a lengthy struggle the British pilot fought his way out of the cockpit and jumped clear of the stricken aircraft.

After their initial firing pass, the Hurricanes split into sections and curved round to attempt follow-up attacks on the bombers from their flanks and from astern. Repeatedly the escorting Messerschmitts dived on the would-be assailants to break up their attacks. For the German fighters assigned to the close escort force it was a frustrating business. They could not pursue the enemy fighters to the kill, for to do so would have meant leaving their charges, and that was strictly forbidden. Again and again the Messerschmitts had to break off the chase and return to their bombers, only to see the British fighters return. Then the whole process had to be repeated.

As the raiders neared Gravesend the British fighter attacks slackened off, and ahead the bomber crews could see the distant smudge of London on the horizon. But now the raiders faced a new hazard. Defending the south and west of Chatham was a concentration of twenty 4.5in and eight 3.7in guns which now opened up a heavy cannonade. *Oberleutnant* Peter Schierning, the navigator in a Heinkel of *Kampfgeschwader 53*, saw a series of black lumps suddenly appear in the sky around him. They came from exploding anti-aircraft shells, some of them so close that their blast rocked the bomber. Within seconds the smoke puffs had been left behind, but by then the damage had been done. Schierning recalled:

> One of the first salvos knocked out our right motor. We felt no shock, but the motor slowly wound down. The pilot shouted 'Get rid of the bombs! Get rid of the bombs! I can't hold it!' I jettisoned the bombs over farmland.

Unable to keep up with the formation, the bomber entered a steep diving turn, making for the protection of a bank of cloud. Another German bomber suffered a fate similar and it too was forced out of its formation and made a dash for the cloud bank.

As during the earlier action, Keith Park again concentrated the bulk of his fighters immediately in front of London for the main engagement. No fewer than nineteen squadrons of fighters, with 185 Spitfires and Hurricanes, were now moving into blocking positions to the south and the east of the city.

As Douglas Bader's 'Big Wing' reached the capital it came under attack from free-hunting Bf 109s diving from above. The three Hurricane squadrons split up to fight off their tormentors and, in a reversal of their usual role, Bader ordered the two Spitfire squadrons to try to get through to the bombers alone. Yet although the 'Big Wing' was unable to deliver the intended massed attack on the enemy bombers, by its presence it kept the enemy's free-hunting patrols occupied and made it easier for other British squadrons to reach the bombers.

The No 11 Group squadrons waiting in front of the capital went into action as the bombers came into sight. Flying Officer Tom Neil of No 249 Squadron took part in the attack on one of the formations of Dorniers and in his book *Gun Button to Fire* later wrote:

> Closing, I fired immediately and the whole of the port side of the German aircraft was engulfed in my tracer. The effect was instantaneous; there was a splash of something like water being struck with the back of a spoon. Beside myself with excitement, I fired again, a longish burst, and finding that I was too close, fell back a little but kept my position. Then, astonishingly, before I was ready to renew my assault, two large objects detached themselves from the fuselage and came in my direction, so quickly, in fact, that I had no time to evade. Comprehension barely keeping pace with events, I suddenly recognised spreadeagled arms and legs as two bodies flew past my head, heavy with the bulges that were undeveloped parachutes. The crew! Baling out! I veered away, shocked by what I had just achieved.

Almost certainly the bomber that Neil hit was that carrying the *Kampfgeschwader 3* formation leader, *Hauptmann* Ernst Püttmann, and the aircraft crashed into the Thames.

During this series of attacks three other Dorniers were knocked out of Püttmann's formation. *Oberleutnant* Hans Schmoller-Haldy of *Jagdgeschwader 54*, flying as close escort with the formation of Heinkels coming up behind, watched the planes go down:

There were parachutes all over the place. Several British fighters were buzzing around the Dorniers. I thought 'Oh, those poor men . . .' But we couldn't do anything to help; we had to stay with our Heinkels.

On the way to the target four German bombers had been shot down, and seven were forced out of formation and heading for home alone. All five of the bomber formations reached London intact, however, and now they were prepared to commence their bombing runs on their assigned dock targets.

Throughout the morning and the early afternoon, cloud had gradually been thickening over southern England. Now most of the capital lay under nine-tenths cumulus and strato-cumulus cloud, with the base at about 2,000ft and tops extending to 12,000ft. And, as luck would have it, none of the targets the German crews had been briefed to attack was visible. West Ham was one of the few parts of the metropolis clear of cloud, and two formations of Heinkels and one of Dorniers re-aligned their attack runs to hit the borough and the nearby gasworks at Bromley-by-Bow. Their bombs caused widespread damage throughout the area.

On the southern flank of the German force, the two formations of Dorniers from *Kampfgeschwader 2*, briefed to attack the Surrey Commercial Docks, arrived to find no part of the complex discernible. The raiders turned away without bombing, much to the surprise and delight of the three squadrons of Hurricanes that had been battling with the Dorniers and their escorts. Many of the RAF pilots thought that by their presence they had scared the enemy into making the U-turn, and this version would be given wide prominence later in press reports. In fact both Dornier formations arrived over the capital intact, having lost only one aircraft on the way in; they could easily have attacked their briefed targets if their crews could only have seen them. On their way home the raiders bombed on 'targets of opportunity', causing scattered damage in the boroughs of Penge, Bexley, Crayford, Dartford and Orpington.

Those German bombers that had been forced out of formations engaged in a deadly game of hide-and-seek, as they tried to pick their way through the banks of cloud over Kent to avoid the defending fighters. Peter Schierning's Heinkel was one of those bombers, limping home with one engine wrecked by flak:

By the engine, part of the skin of the wing had been blown away and I could see the structure inside. I remember thinking what a marvellous aircraft the Heinkel was, being able to stay airborne with that sort of damage.

Then, suddenly, both the cloud and the bomber crew's luck ran out. Surrounded by clear skies, almost at once the Heinkel came under attack from a couple of Hurricanes and a Spitfire. Schierning recalled:

> They attacked from the rear and the sides, and I saw their tracers coming past the nose of the Heinkel. Early on, the intercom was shot away and I had no idea what was happening in the rear of the aircraft; I had no opportunity to use my nose gun.

The Hurricane pilots, from No 229 Squadron, were Squadron Leader A. Banham and Flight Lieutenant E. Smith. The latter reported that with his commander he went after a lone Heinkel that was also being engaged by a Spitfire:

> I attacked from dead astern with a 6-second burst from 200 yards, closing to 150 yards. I saw bullets entering the fuselage and hitting the mainplane. The port engine was smoking and brown oil from the E/A splashed over my windscreen.

When Smith last saw the bomber it was diving for a patch of cloud. Schierning heard his pilot shout that the port engine had also been hit and was losing power, there was a fire in the vicinity of the right engine and he was taking the bomber down for a crash landing. One of the gunners was killed and the radio operator was wounded. The Heinkel crash-landed near Staplehurst and slithered to a halt in a cloud of dust.

Despite the presence of cloud, which made the plotting of aircraft difficult over land, during the afternoon action the RAF system of fighter control surpassed even the morning's fine performance. Twenty-eight squadrons, with 276 Spitfires and Hurricanes, had been scrambled to meet the afternoon attack. Every one of those squadrons went into action.

Although the main attacks of the day were against London, the city was not the only target. In the expectation that Fighter Command would concentrate all of its remaining strength for the defence of the capital, twenty-seven Heinkels of *Kampfgeschwader 55* were sent to make an unescorted attack on the Royal Navy base at Portland. Only a lapse by the British fighter controllers saved the raiders from heavy losses. The strength of the attack force was underestimated, and just one half-squadron of Spitfires was sent to intercept it. Only one bomber was shot down, although another suffered damage. The bombing of the naval installation was inaccurate and it caused no significant damage.

Later in the afternoon ten Messerschmitt Bf 110s and three Bf 109s of the fighter-bomber unit *Erprobungsgruppe 210* ran in at low altitude to attack the Supermarine aircraft works at Southampton, then building Spitfires. Although the raiders reached the port without interference, they failed to find their intended target and the weight of their attack fell on a residential area a few hundred yards to the east. During their withdrawal the fighter-bombers fought a brief skirmish with Hurricanes of No 607 Squadron, but there were no losses on either side.

In the course of the hard-fought series of actions on 15 September 1940, the *Luftwaffe* lost a total of 56 fighters and bombers. RAF Fighter Command lost 29 aircraft. At the time the defenders claimed the destruction of 185 German planes, the largest of many huge overclaims that were made during the Battle of Britain.

A close analysis of the day's actions puts Douglas Bader's 'Big Wing' tactics in a new light. During the Battle of Britain the five-squadron fighting unit from No 12 Group made contact with a large force of enemy aircraft on three occasions, and two of those were on 15 September. At noon on that day the 'Big Wing' went into action on the most favourable terms imaginable, when its 56 fighters engaged a formation of 25 Dorniers with no escorts. An air combat that involves large numbers of aircraft is likely to give rise to large overclaims in the number of enemy planes destroyed. Furthermore, by definition, if the 'Big Wing' took part, then a large number of aircraft were involved. During the noon action the 'Big Wing' claimed 26 German aircraft destroyed, including twenty bombers, for no loss to itself. From German records and interviews with surviving bomber crewmen it is clear that only six bombers were lost during the entire action. It appears likely that fighters of the 'Big Wing' shared in the destruction of five, possibly all six, of the Dorniers shot down; but every one of those bombers was also attacked by fighters of No 11 Group.

During the afternoon action German fighters broke up the 'Big Wing' before it could get close to the bombers. The Wing split into small units and its pilots claimed the destruction of 26 enemy aircraft, including seventeen bombers. From German records it is clear that only 21 bombers were lost during the entire engagement and again, from British records, it is clear that most of those were attacked by fighters of No 11 Group.

The 'Big Wing' actions on September 15 resulted in huge overclaims of the number of German aircraft destroyed, which produced an exaggerated assessment of the effectiveness of these tactics. During the noon action there were cases of fighters getting in each others' way, and attacks had to be broken off to avoid collisions. In terms of enemy aircraft destroyed, the 'Big Wing' was less effective than five squadrons of fighters sent into action in ones and twos.

The failure of the 'Big Wing' to shoot down many bombers was more than counter-balanced, however, by the resounding success that it had in another area—its devastating impact on the morale of the German bomber crews. During their briefings in the morning of 15 September, the raiders had been told that Fighter Command was a spent force that need no longer be feared. The repeated attacks by Spitfires and Hurricanes on the way to the capital cast doubts on the accuracy of that intelligence. But then, as they arrived over London, the raiders were confronted by more than fifty Royal Air Force fighters approaching in parade formation. That display, mounted twice during the day, demonstrated beyond possible doubt that the Fighter Command was far from beaten. By impressing that unpalatable fact on the *Luftwaffe*, the 'Big Wing' was well worth the effort involved.

Because of the huge victory claim, which we now know to be exaggerated, 15 September has come to be commemorated as 'Battle of Britain Day'. In fact that date did mark the decisive point in the Battle, though for a different reason. The strength of Fighter Command's reaction during the two actions around London convinced Hitler that the *Luftwaffe* could not gain air superiority before the weather broke that autumn. On 17 September he ordered that Operation 'Sealion', the planned invasion of southern England, be postponed 'indefinitely'. The ships and barges assembled for the enterprise at ports along the Channel coast began to slip away to resume their normal tasks. With each week that passed the threat of invasion lessened, never to return. Although it was chosen for the wrong reason, the date commemorated by the British public for the deliverance of the nation was the correct one.

3

PRE-EMPTIVE STRIKE

On 22 June 1941, without warning or prior declaration of war, German forces launched a massive attack on the Soviet Union. During the first day one of the primary objectives of the Luftwaffe *was the destruction of the Soviet Air Force. The latter was quite unprepared to meet the blow and, as a result, it suffered the heaviest defeat ever experienced by an air arm.*

DESPITE THE non-aggression pact signed between the governments of Germany and the Soviet Union in 1939, there was little trust between the two. The treaty gave the Soviet government a free hand in the east. Its army occupied Latvia, Estonia and Lithuania and parts of Poland, Rumania and Finland, and the additional territories were incorporated into the Soviet Union. Moreover, between the summer of 1939 and May 1941 the Soviet Army increased in strength from 69 to 158 divisions, the majority of them positioned along the nation's western frontiers. During the same period the Soviet Air Force also expanded and, significantly, at the end of it many units were about to re-equip with the high-performance new bomber and fighter types then coming off the production lines. Adolf Hitler believed that, sooner or later, Germany would find herself at war with the powerful communist state to the east, and he resolved to get his blow in first.

During the spring of 1941, while the night blitz on Great Britain was in full swing, *Luftwaffe* combat units in western Europe secretly moved to bases in eastern Germany and Poland. The transfer was painstakingly planned to keep its true intent a secret for as long as possible. *Leutnant* Dieter Lukesch, a Junkers Ju 88 pilot who served as technical officer with *III Gruppe* of *Kampfgeschwader 76* based at Cormeilles-en-Vexin near Paris, described the elaborate subterfuge. The first indication he had that something strange was afoot was when his unit was ordered to remove the temporary black distemper applied to the aircraft for night operations and repaint the upper surfaces in light brown camouflage. That suggested daylight operations in a desert area, but where? Half way through the work the order was countermanded, then a new order arrived to the effect that the planes were to be restored to their original temperate colour scheme, with the topsides camouflaged in two shades of green. When this was done, the bulk of the unit's technical personnel suddenly departed for an undisclosed destination, leaving behind a few men to look after the aircraft.

Rumour followed counter-rumour, and a few days later the mystery deepened, as Lukesch explained:

The air crew were summoned to a meeting in the middle of the airfield, well clear of everyone else. There the *Gruppe* commander, *Major* Lindmayr, solemnly opened an envelope that contained our sealed orders. What followed only served to heighten our curiosity. Our planes were fuelled up. We were told to load our personal kit on the aircraft, then take off and form up by *Staffeln* behind Lindmayr, who was to lead us to our still-secret destination. We took off from Cormeilles and flew over Holland and Germany before landing at Anklam [a *Luftwaffe* airfield on the Baltic coast]. After we taxied in and shut down the planes, we were driven to a barrack block where we were kept in isolation. Everything there had been prepared for us, our beds were made, the tables had been laid and a meal was ready.

The following morning there was a near-repeat of the previous performance. Again there was the briefing on the airfield, the brown envelope was solemnly opened and again the crews were told to take off and follow Lindmayr to the undisclosed destination. Lukesch continued:

This time, after a flight of two hours, we landed at Schippanbei just south of Königsberg. When we arrived we found that our technical people were already there; they marshalled us into prepared camouflaged dispersal points around the airfield. The aircraft were then carefully concealed under camouflage netting and branches cut from trees. Then the planes were refuelled and bombed up, but still we did not know where we were going.

For *II Gruppe* of *Kampfgeschwader 3* it was a similar story, though this unit started its move from Oldenburg in Germany, where it had been converting from Dornier Do 17s to Ju 88s. The *Gruppe* moved to Podlotowka near Brest-Litovsk in Poland, and one of the pilots, *Feldwebel* Horst Schulz, recalled:

When we arrived at Podlotowka we saw a lot of army units there, infantry, artillery and tanks. Rumours were rife and the most popular was that the Russians were going to let a German force of two or three divisions with air support through their territory to attack the British oil fields and pipelines in Iran.

During the afternoon and evening of 21 June the men finally learned whom they were to attack, and when. The new enemy was the Soviet Union, and the war would begin with a series of co-ordinated land and air attacks, to commence simultaneously before dawn the following morning. The men were assembled and each commander read out an personal order of the day from Adolf Hitler that began 'Soldiers of the Eastern Front . . .' The missive stated that, despite the treaty of friendship between the two nations, German intelligence had discovered that Soviet forces were massing for a treacherous onslaught against Germany. As a result, the *Führer* had reluctantly ordered the counter-stroke in order to blunt the force of the planned attack before it could begin.

At 3.00 a.m. Central European Time on 22 June the German Army opened its offensive with a massive artillery bombardment all along the front, in true *Blitzkrieg* style. Then the spearhead units began moving forwards. Between the Baltic and the Black Sea, 117 German divisions, of which 48 were armoured, plus fourteen Rumanian divisions and a Hungarian army corps, swung into action. Facing them in the immediate battle area were 132 Soviet Army divisions, of which 34 were armoured. A total of 7¾ million men were thus committed in one of the greatest armed clashes in history.

On the first day of the campaign the *Luftwaffe* had a total of 1,954 combat aircraft deployed to support the attack on the Soviet Union, of which just over two-thirds were serviceable: 510 twin-engine bombers, 290 dive-bombers, 440 single-engine fighters, 40 twin-engine fighters and 120 reconnaissance planes. The earliest that the *Luftwaffe* could attack targets in force was about half an hour after sunrise, or 4.20 a.m. This would enable aircraft to take off and assemble into formation in the morning twilight and reach their targets to strike with greatest effect. German Army commanders decreed that the earlier starting time was necessary, so that their own attacks could achieve surprise, however. The best that the air force could do was to send small numbers of Heinkel He 111s, Dornier Do 17s and Ju 88s flown by picked crews, in three-plane units with each machine flying independently in the darkness, to attack some of the more important Soviet fighter airfields. These attacks took place simultaneously at 3.15 a.m., fifteen minutes after zero hour. Their purpose was to disrupt operations at the airfields, and in particular to delay the dispersal of aircraft on the ground until the arrival of the large-scale air attacks after dawn.

As the daylight attack units soon discovered, however, the state of readiness of the Soviet units was so poor that, even at airfields that had not been attacked, no measures had been taken to meet such a possibility. Long conditioned to obey those above without question, the Soviet airfield commanders were afraid to do anything until they received orders. Had they actually wanted to make things easy for their attackers, it is difficult to see what more the Soviets could have done to assist in the destruction of their planes. Scores of fighters, bombers and reconnaissance aircraft sat on the ground in neat rows at their airfields. Putting the planes close together meant that a single bomb could damage to two or more, and putting them in a line meant that a strafing fighter or bomber could fly down the line at low altitude and fire at each one in turn. Even the simple precaution of dispersing the aircraft around the perimeter of each airfield, with a hundred yards or so between each, would have made the raiders' task more difficult and their attack less effective. Taxying the planes a short distance off the airfield and applying simple camouflage would have made things more difficult still. And a few anti-aircraft guns around each airfield would have served to distract the attackers at the critical time and make their bombing runs less accurate. But, so confident were the Soviet leaders that no attack was in the offing, none of these things was done. The Soviet Air Force was to suffer accordingly.

It was light when Dieter Lukesch and his *Gruppe* took off to attack the airfield at Krudziai south of Riga in Lithuania. Although he already had flown several combat missions against Great Britain, this was one he would never forget:

> The skies were beautifully clear, with visibility almost unlimited. Soon after we took off we could see the front line quite clearly, marked by fires and the smoke from bursting shells. Once we had passed the front, however, there was no flak. We did not have, nor did we expect to need, an escort for the first attack. As we passed other airfields we saw

Russian fighters taking off, but they climbed somewhat slower than our cruising speed so we soon left them behind.

The Ju 88s cruised at 10,000ft, each carrying the standard load of four 550lb and ten 110lb general-purpose bombs. Five miles short of the target the planes dropped their noses and began shallow-dive attacks to plant their bombs with the greatest possible accuracy. The attack opened at 6 a.m., but although it was some hours after the opening bombardment more than a score of the Tupolev SB-2 bombers were drawn up in line along one side of the airfield. Lukesch continued:

There was no flak, and even though the war had been in progress for about three hours it seems that we had achieved surprise. As we approached for the first attack we could see ground crewmen standing on the wings refuelling the aircraft, looking up in curiosity as we ran in. As the bombs started to explode they made hasty retreats into the nearby forest. Our bombers attacked in a long line but there was some jockeying for position. As I was about to release my bombs I saw another aircraft converging on me from the right and releasing its bombs. I had to break away, make a circuit and attack at the end of the force. I ran in as the last aircraft in the *Gruppe* to attack. By then several aircraft on the ground were burning and there was quite a lot of smoke, but the line of trees behind the aircraft helped me to line up on some planes that had not been hit before. During the attack my observer fired at the enemy planes with his machine gun. As we were pulled away after the attack some Russian fighters appeared on the scene, Ratas and Gulls [Polikarpov I-16s and I-15s]. Although they got close the did not fire at us; perhaps they did not have any ammunition. With my greater speed I soon left them behind.

That morning some *Luftwaffe* units employed a new weapon in action for the first time: the 4½lb SD-2 fragmentation bomb (sometimes called the 'Butterfly bomb'), carried in special containers fixed to the attacking planes. After release, the casing of the SD-2 opened up to form a pair of 'wings' and the weapon spun to the ground like a sycamore seed. The 7oz explosive charge detonated on impact, hurling high-velocity fragments in all directions with sufficient force to cause serious damage to aircraft up to 40ft away. Dropped in large numbers during low-flying attacks on Soviet airfields, the SD-2s proved highly effective against aircraft and other 'soft' targets. *II Gruppe* of *Jagdgeschwader 27* was one of the units to employ the new weapon, its Bf 109E fighters each carrying 96 of the small bombs. That morning it sent 31 fighters to deliver a low-altitude bombing and strafing attack on Wizna airfield. By the end of this attack and one by fifteen dive-bombers, a total of 31 Soviet aircraft had been destroyed. The two raiding forces then went on to attack Lomza-South airfield, where they destroyed another 40 planes. All of the fighters from *JG 27* returned safely.

Kampfgeschwader 51 had a much less happy experience with the new weapon. That morning the *Geschwader* dispatched all 91 serviceable Ju 88s in attacks on six major Soviet airfields on the southern part of the front, each plane carrying 360 SD-2s. The bombing and strafing attack on Stryj airfield, by eighteen Ju 88s, caused the destruction of twenty Soviet bombers; the raiders then continued to Lemberg airfield, where they destroyed fifteen fighters. After the attack crews learned that the new fragmentation bombs could also be extremely dangerous companions, however. Bombs were liable to jam in the container, and the plane's crew had no way of knowing of the 'hang-up'. Because of a design fault, the bomb's fuse could become live in flight and thereafter the slightest shock might set off the charge—with disastrous results for the aircraft; alternatively, on landing, a jammed SD-2 was liable to jolt free, fall to the ground and explode. On the first day of the campaign *KG 51* lost fifteen bombers to enemy action or to accidents with SD-2s—nearly half the number of

POLIKARPOV I-16 TYPE 24

Role: Single-seat fighter.
Powerplant: One Shvetsov M-62 9-cylinder, air-cooled, radial engine developing 1,000hp for take-off.
Armament: Two ShVAK 20mm cannon; two ShKAS 7.62mm machine guns.
Performance: Maximum speed 326mph at sea level, 286mph at 14,750ft; climb to 16,400ft, 5min 48sec.
Normal operational take-off weight: 4,215lb.
Dimensions: Span 29ft 6½in; length 20ft 1¼in; wing area 161 sq ft.
Date of first production I-16 Type 24: Summer 1939.

aircraft lost by the entire *Luftwaffe* on that day—and the SD-2s immediately gained the grim nickname 'Devil's Eggs'.

While the attacks were in progress against Soviet airfields, other *Luftwaffe* units attacked targets in support of advancing Germany ground troops. Dive-bombers attacked fortified positions, head-quarters, artillery parks and barracks. Horst Schultz's *Gruppe, II* of *KG 3*, sent out several Ju 88s on single-plane armed reconnaissance missions along roads in the Soviet rear areas leading to the battle front. He recalled:

> We went free hunting on our first mission [over the USSR], flying at 1,200m [about 3,800ft] looking for enemy road traffic. We dropped our bombs on a road filled with troops and artillery, and on a bridge over which troops were passing.

Simultaneously, packs of German fighters ranged over the battle area hunting for any Soviet planes that had got into the air. Again, it was a one-sided battle. Close to the ground the Polikarpov I-16 Type 24, the main Soviet fighter type, was almost as fast as the Bf 109F that equipped the majority of *Luftwaffe* fighter units. But the Type 24's radial engine was optimized for low-altitude operations, and as height increased its performance fell away steadily: at 20,000ft the I-16 was about 100mph slower than the German fighter. The Soviet fighters were more manoeuvrable than their adversaries, but in a fighter-versus-fighter combat this advantage did no more than allow a pilot to avoid being shot down, provided he saw his attacker in good time. The faster fighter always held the initiative, and its pilot dictated the terms of the engagement and could attack or break off the combat at will. For their part, the Soviet fighter pilots had to dance to their enemies' tune.

Typical of the scrappy actions taking place that morning was one near Brest-Litovsk, described by *Unteroffizier* Reibel of *I Gruppe* of *Jagdgeschwader 53*:

> I was flying as wing man to *Lt* Zellot. We flew in the direction of Brest from Labinka. As my leader ordered a turn about, I saw two biplanes in front of us. I immediately reported them and we brought them under attack. When we were about 200m from them they both pulled into a tight turn to the right. We pulled up high and then began a new attack, but though we both opened fire it was without success. Soon there were about ten other [enemy] machines in the area. My leader ran in to attack one of the planes while I remained high in order to cover him. Then an I-15 became separated from the others. I immediately prepared to attack it, but I had to break away when another enemy machine, which I had

TUPOLEV SB-2bis

Role: Three-seat medium bomber.
Powerplant: Two M-103 inline engines each developing 960hp at take-off.
Armament: Normal operational bomb load 2,200lb. Defensive armament of five 7.62mm machine guns, two in the fuselage gun turret, two on a flexible mounting in the nose and one firing downwards and rearwards from the ventral gun position.
Performance: Maximum speed 280mph at 12,900ft; radius of action with 2,200lb bomb load, 200 miles.
Normal operational take-off weight: 10,500lb.
Dimensions: Span 66ft 8½in; length 41ft 2½in; wing area 610 sq ft.
Date of first production SB-2bis: 1937.

not seen, suddenly appeared 50 metres in front of me. I opened fire with machine guns and the cannon and it burst into flames and spun out of control. Apparently the pilot had bailed out. Then I had to turn away, as I had two [enemy] machines behind me.

Using their superior speed, the German fighter pilots easily pulled clear of their pursuers.

JG 27 lost its commander, *Major* Wolfgang Schellmann, near Grodno when he pressed home an attack on an I-16 to short range. Following an accurate burst the Soviet fighter exploded and debris from the aircraft struck Schellmann's fighter. He was forced to bail out and was taken prisoner. It appears that he was shot soon afterwards by his captors.

German twin-engine bombers and dive-bombers had flown a total of 637 sorties by 10 a.m. that morning, striking at 31 airfields as well as military positions and supply routes. The success of the attacks on the Soviet Air Force that morning is confirmed by the official Soviet postwar publication *History of the Great Patriotic War of the Soviet Union* (not a source likely to exaggerate German successes):

During the first days of the war enemy bomber formations launched massive attacks on sixty-six airfields in the frontier region, and in particular those where the new Soviet fighter types were based. The result of these raids and the violent air-to-air battles was a loss to us, as at noon on 22nd June, of some 1,200 aircraft, including more than 800 destroyed on the ground.

From mid-day small forces of Soviet bombers attempted to deliver retaliatory attacks on airfields used by the *Luftwaffe*, but with little success. About a dozen Tupolev SB-2s carried out a high-level bombing attack on the airfield at Biala-Podlaska just inside German-held

Poland, home of the Junkers Ju 87 dive-bombers of *I Gruppe* of *Sturzkampfgeschwader 77*. The Stukas were being refuelled and re-armed after their initial mission but, in contrast to those of their enemy earlier in the day, the German planes were dispersed around the airfield and camouflaged. A flak battery positioned nearby went into action against the raiders, and although some of the bombs burst across the airfield, no Stukas were damaged.

Summoned to the scene by the flak bursts, German fighters were soon converging on the Tupolevs from several directions. One Bf 109F pilot who took part in the action, *Oberleutnant* Ohly of *I Gruppe* of *JG 53*, later reported that he had scrambled with four fighters at 12.18 following a report of enemy planes in the Brest-Litovsk area heading west:

> About [8,000ft] over Biala Polkaska, I saw flak bursts and a formation of about a dozen multi-engined monoplanes that turned out to be SB-2s. During the climb I lost contact with the two other pilots, so I headed towards the SB-2s with just our a pair. These [enemy] aircraft had already been engaged by other fighters and were heading towards the Bug [river]. I engaged three SB-2s flying in vic formation, and fired at the machine on the right with my cannon and machine guns from a range of between 150 and 75 metres. My speed was too great and I had to pull up [to avoid colliding with] the SB-2, and as I flew past it my aircraft received hits in the radiator and the fuselage. I made a left turn to check that my aircraft responded to the controls, during which the machine I had attacked passed out of view. I turned through 360 degrees and the SB-2s again came into view. I watched my wing-man and another fighter shoot down three SB-2s. At the same time several SB-2s that had been attacked by other fighters were also going down. I cannot say if the one that I had hit was shot down.

The Soviet bomber formation lost about three quarters of its aircraft during the series of attacks by German fighters. As a footnote to the day's air fighting, there were five cases reported in which Soviet fighters deliberately rammed enemy planes. This practice often resulted in the death of all of those aboard both planes, and it was a grisly portent of what lay in store for the *Luftwaffe* when the fast-moving *Blitzkrieg* campaign drifted into a long war of attrition.

By the end of the first day of its campaign against the Soviet Union the *Luftwaffe* had flown a total of 1,766 sorties by single-engine and twin-engine bombers and 506 by fighters. In the course of these operations it lost thirty-five aircraft. Official German sources claimed the destruction of 322 Soviet planes in air-to-air combat or by flak, and 1,489 destroyed on the ground. Given the size of the aircraft losses admitted by the Soviets up to noon on that fateful day, the German claims undoubtedly have a ring of truth. Certainly the day was one for the aviation record books. In the course of an eighteen-hour period between 3.15 a.m. and sunset the Soviet Air Force losses represented by far the greatest number of aircraft ever destroyed in a single day's fighting. It was also the most comprehensive defeat to be inflicted by one air force on another, in the long history of air warfare. Although five-sixths of the Soviet aircraft were destroyed on the ground, one-sixth, or 322 machines, fell in air-to-air combat; this was the largest number of planes ever to be shot down in a single day.

During the days that followed, the German armoured units thrust rapidly into the Soviet Union, overrunning every one of the airfields that the *Luftwaffe* had attacked on the first day. This compounded the effect of the earlier losses, for aircraft that were not flyable because of battle damage or other types of unserviceability either had to be destroyed by the retreating forces or were left behind to be captured. Having one's ground forces capture enemy airfields is a most effective means of reducing the capability of an enemy air force and one that is often neglected when considering this aspect of warfare.

Yet, despite the enormity of the material losses suffered by the Soviet Air Force during the early days of the war, their effect was crippling only in the short and medium terms. In June 1941 the programme to re-equip the Soviet front-line units with modern aircraft had scarcely begun, with the result that the great majority of the aircraft lost were obsolescent types that were scheduled to be replaced in the near future. Moreover, very few Soviet air crew were lost in the attacks on the airfields or when the latter were captured, so as the modern planes came off the production lines there was no shortage of trained personnel to put them into action. Despite the ferocity of their initial onslaught, the German forces were unable to secure victory in the Eastern Front within the expected five months; at the end of that time they had to face the much improved Soviet ground forces and a revitalized Soviet Air Force. It was not going to be a short war.

4

GUIDED MISSILES MAKE THEIR MARK (1)

The term 'guided missile' can be defined as 'a missile whose path can be corrected during its travel, either automatically or by remote control, to bring it into contact with the target'. Note that this definition makes no mention of the medium through which the weapon usually passes, nor of its speed. To be sure, most such weapons travel through the air at speeds of several hundreds of miles per hour. But the first air-launched guided missile to be used in action was a quite different animal—an anti-submarine torpedo with an underwater running speed of 12kts. During the Second World War the weapon was a closely guarded secret, as were its successes.

URING 1942 scientists and technicians in the United States, Britain and Canada worked on several detection and location devices, and weapons, to counter the U-boats. One weapon that resulted from the US Navy's work in this field was the first air-dropped anti-submarine homing torpedo, a weapon so secret that it was given the deliberately misleading code-name 'Mark 24 Mine'. Although the weapon had a running speed of only 12kts, within most definitions of the term it was a self-homing guided missile.

The diameter of the weapon, 19in, was about the same as that of the standard anti-ship torpedo; however, its 7ft length was less than half that of the standard weapon, and this gave the Mark 24 Mine a disproportionately dumpy appearance. Intended for use against U-boats that were running underwater or that had recently crash-dived, the homing torpedo weighed 640lb, of which 92lb was the high-explosive warhead with an impact fuse in the nose.

The method of operation of the Mark 24 Mine was as follows. Aimed at the diving swirl left by the boat when it disappeared under the waves, or on the most accurate datum point available on the location of the submarine, the weapon was released from an altitude of about 250ft. On entering the water the electric motor started and the torpedo began its search pattern, running round the circumference of a circle about 4,000yds in diameter. The acoustic homing head was tuned to listen for distinctive sounds produced by cavitation, the noise of the popping of bubbles caused when a propeller rotates in water at high speed. When it picked up these sounds, the homing head assessed the bearing and elevation of their source and steered the weapon in that direction until it impacted. If for any reason the acoustic head ceased to pick up sounds on which it could home, the torpedo resumed its search pattern. The weapon had a maximum running time of ten minutes, after which it sank to the bottom of the sea to prevent its capture.

At the end of 1942 the Mark 24 Mine entered series production, and thus became the first self-homing guided missile in the world to achieve this status. Tests revealed that under ideal conditions, with a calm sea and the submarine running at speed just under the surface, the weapon could home on its target from a distance of three-quarters of a mile. However, after a crash-dive it was likely that a boat would be running at speed, because the commander would want to get clear of the tell-take diving swirl as rapidly as possible. Thus while the homing torpedo promised to be highly effective against an unsuspecting foe, the weakness of its concept would be obvious to an enemy who knew its method of operation. The acoustic head could home only on a *cavitating* propeller, so the submarine would be safe if its commander overcame his instincts and slowed the boat to below cavitation speed as soon as it had submerged. If the existence of the homing torpedo became known to the enemy it would be useless, so exaggerated security precautions surrounded every aspect of the weapon. To reduce the possibility of capture, it was not to be used close to enemy-held shores; furthermore, it was not to be used in areas where an attack might be observed by the enemy and its method of operation deduced.

An old adage assures us that 'It is almost impossible to make a system completely foolproof, because fools are so resourceful', and the Mark 24 Mine nearly fell foul of this logic a few months before it was ready for action. In February 1943 an early-production homing torpedo was transported to Great Britain aboard the liner *Empress of Scotland*. Group Captain Jeaff Greswell, returning from a liaison visit to the United States, was the custodian of the top-secret weapon, and he described the elaborate security precautions attending the move:

> The homing torpedo arrived at the quay at New York in a US Navy lorry escorted by armed guards; there seemed to be guns all over the place. The weapon was packed in three large boxes: one contained the nose, one the centre section and one the tail. These the sailors brought on to the ship, they formally handed them over to me and I signed for them. Then I witnessed the placing of the boxes in the captain's safe, and he gave me a receipt. At Liverpool it was the same thing in reverse. There we were met any an RAF lorry, again with armed guards. I received the boxes from the Captain and signed for them, then I handed them over to the

RAF officer and obtained his receipt. The demands of security had been observed to the letter. Now my part in the operation was over and I went off for a few days' leave. I had been at home for a couple of days when I received a buff-coloured envelope with OHMS across the top, by ordinary post. Inside was a letter from His Majesty's Customs; they wanted to know why I had imported into the United Kingdom 'packing cases containing what is believed to be some form of aerial homing torpedo for use against submarines'. Why had I failed to declare them?

Greswell passed the letter with a secret covering note to Air Chief Marshal Sir Philip Joubert, Commander-in-Chief of Coastal Command, with a frantic appeal: 'For heaven's sake do something about this one.' The Group Captain heard no more from the bureaucrats.

By May the homing torpedo was ready to go into action. Among the first units to receive it were US Navy Patrol Squadron VP-84, equipped with Catalina amphibians based in Iceland, and two Royal Air Force Liberator units, No 86 Squadron based at Aldergrove in Northern Ireland and No 120 Squadron based at Reykjavik in Iceland. All three squadrons were engaged in flying anti-U-boat patrols in support of convoys in mid-Atlantic.

Early in May the twin eastbound convoys HX.237 and SC.129 set sail from Newfoundland bound for Liverpool, and from 'Ultra' decoded signals Royal Navy Intelligence officers learned that a German wolfpack with no fewer than 36 U-boats was moving into position to engage the merchantmen. By the morning of the 12th the U-boats were closing on their prey and three Liberators from No 86 Squadron were sent to patrol ahead of the convoys, with the aim of driving the U-boats underwater to prevent them from reaching their attack positions. Each of these aircraft was loaded with four 250lb depth charges and two Mark 24 Mines. If an aircraft found a U-boat and it remained on the surface, they were to attack it with depth charges; if the boat submerged, the crew would attack it with a homing torpedo.

The honour of being the first to use a guided missile in action goes to Flight Lieutenant J. Wright and his Liberator crew. They located *U456* on the surface, and as they closed in to deliver their attack the boat obligingly dived. Wright took the aircraft low over the diving swirl and released a homing torpedo, together with a smoke float to mark the point of entry. Then he swung the aircraft round in a wide circle, while he and his crew waited for some indication that the new weapon had been effective. For two long minutes nothing happened then, about 900yds from the point where the torpedo had entered the water, there was a small upheaval as though a depth charge had exploded with less than its normal force (the homing torpedo's warhead contained less than half the explosive charge of an air-dropped depth charge). Shortly afterwards the damaged U-boat re-surfaced, but by then the Liberator was close to the limit of its endurance and Wright was forced to head for base. From German records it is known that the U-boat reported by radio that it had suffered damage in an air attack and was unable to dive, and it requested assistance. The boat survived the night, but the next day a couple of the convoy escorts finished her off.

The action around the two convoys continued, and on the 14th aircraft delivered attacks with homing torpedoes on two occasions. In separate actions against two U-boats that had dived as aircraft approached, a Liberator of No 86 Squadron and a Catalina of VP-84 each released a homing torpedo beside the diving swirl. In each case, more than a minute later, a mushroom-like disturbance appeared in the water some distance from the entry point. No other results were observed, but from German records it is known that *U226* and *U657* disappeared without trace in positions which corresponded to these attacks and the homing torpedo is credited with their destruction. On the 19th, a Liberator of No 120 Squadron used one of the new weapons to dispatch *U954*.

From then on the Mark 24 Mine, affectionately nicknamed 'Wandering Annie' by those who used it, took an increasing toll of the

German submarines. During the early summer of 1943 the US Navy escort carrier *Santee* took part in a series of actions off the Azores, and developed an effective technique against U-boats that stayed on the surface and used their anti-aircraft weapons to try to fight off their assailants. The carrier dispatched pairs of aircraft during U-boat hunts, a Grumman F4F Wildcat fighter teamed with a Grumman TBF Avenger bomber. The Avenger used its radar to locate the U-boat running on the surface and directed the Wildcat into position to deliver a strafing attack to force the submarine to dive. Then the Avenger planted a homing torpedo beside the diving swirl. Using these tactics *Santee*'s aircraft sank three U-boats during July 1943.

During 1944 the Coastal Command Tactical Development Unit put together a elaborate procedure for using the Mark 24 Mine to attack U-boats that had been submerged for some time or those whose diving swirl was not visible (i.e. in poor weather or at night). First the aircraft laid out a pattern of five sonobuoys, each with a smoke marker (at night a flame float) to mark its position. By listening to the underwater sounds transmitted from each buoy in turn, an operator in the aircraft made a 'guesstimate' of the U-boat's position. Then the aircraft ran in and released one or more homing torpedoes at the position thus found. The tactics pushed the early non-directional sonobuoys and the Mark 24 Mine to the limits of their capabilities, and to succeed they required a high degree of skill and teamwork from the crew of the aircraft, plus a measure of luck.

The technique brought its first success on 20 March 1945. 'Ultra' decrypts had indicated that a U-boat on the way to its patrol area would pass close to the Orkney Islands that evening, and a Liberator of No 86 Squadron was dispatched to hunt for it. It was nearly dark when Flight Lieutenant N. Smith and his crew began their search, and soon afterwards the plane's radar operator reported a suspicious contact at a range of three miles. Smith turned towards the object, but as the Liberator closed to within half a mile the object disappeared into sea clutter. In the darkness the plane's look-outs saw nothing of the contact and at this stage it did not justify the expenditure of expensive weapons: the object might have been nothing more significant than a piece of flotsam.

Smith's suspicions had been aroused, however, and he decided to put down a pattern of sonobuoys to see if there was something less innocuous there. The first sonobuoy was released at the point where the object had been seen on radar, to serve as the centre of the pattern. This was followed by four more buoys, laid out in a square with each one at a distance of 2¼ miles from the central buoy. A flame float was released with each sonobuoy, to mark its position.

Laying out a sonobuoy pattern was a complicated process requiring very precise flying, and it took the crew thirteen minutes to position all five buoys in the water. But when the first sonobuoy came on the air they immediately had their reward: the operator heard the unmistakable *swish* of a cavitating propeller rotating at 114rpm—a submerged U-boat. One by one the other sonobuoys began transmitting, and the operator was able to narrow the position of the boat to one part of the pattern. As he was doing so, the radar operator caught another short glimpse of the object he had first seen. It was now clear that the mystery object was the *Schnorkel* head of a submarine that was running just below the surface. With the target identified and located sufficiently for his purpose, Smith took the Liberator to the opposite side of the pattern and began his attack run. As the aircraft passed over the flame float in the centre of the pattern, the navigator began counting off the seconds for a timed run along the bearing where the U-boat was thought to be. When the aircraft reached the boat's computed position Smith released one homing torpedo and, after a measured interval, a second.

The painstaking aerial activity over that patch of water had lasted for more than 20 minutes, during which the unsuspecting U-boat crew continued towards their operating area. It took a further six minutes for one of the homing torpedoes to catch up with the boat, then the sonobuoy operator in the Liberator heard a long, reverberating sound in his phones followed only by sea noises. At the time that was the sole evidence of a successful attack, but from German records it is clear that *U905* had disappeared in that area at about the time of Smith's attack.

Between May 1943 and the end of the war a total of 346 Mark 24 Mines were dropped in action, and the weapon was credited with the destruction of 38 submarines and with causing damage to a further 33. The first air-launched guided missile to be used in action, it was also the most successful weapon of this type to be used during the Second World War: under operational conditions about 20 per cent of the weapons used scored hits on targets.

5

GUIDED MISSILES MAKE THEIR MARK (2)

In the previous chapter we observed the development and use of the first 'guided missile' to go into action. Within a few months two other types of guided missile saw combat, weapons that were quite different from each other and from the one employed previously. This chapter describes the Luftwaffe *operations against Allied shipping using its new guided weapons.*

URING THE Second World War at sea Germany faced two of the world's largest maritime powers, each of which possessed a fleet much larger than her own. In order to redress this imbalance in naval forces, much would depend on the ability of the *Luftwaffe* to deliver effective attacks on naval targets. During the early part of the war Allied ships had weak anti-aircraft defences and, as a result, aircraft could deliver accurate attacks from low altitude using con- ventional high-explosive bombs. That phase soon passed, however, and air crews learned that if they were to survive they had to employ less risky modes of attack. High-level bombing was generally in- effective against warships manoeuvring at speed in open water. Dive-bombing was effective against such targets, but it required clear skies to 10,000ft and the short radius of action of the Junkers Ju 87 Stuka precluded it from attacking targets far from the coast. German torpedo bombers scored some successes, but regular training was necessary to keep crews proficient, the force could not be employed on other types of war mission and its opportunities to go into action were few and far between.

Because of these limitations, the *Luftwaffe* accorded the highest priority to the development of more effective weapons for use against enemy shipping. In November 1941, over the Baltic, flight trials began of a completely new radio-controlled guided weapon, the Henschel Hs 293 glider bomb. Intended for attacks on unarmoured warships or transports, the Hs 293 resembled a small winged aircraft and carried a 1,100lb warhead in the nose. Following its release from the launching aircraft, a rocket motor accelerated the weapon to a maximum speed of about 375mph. After a burning time of twelve seconds the rocket cut out and the missile then coasted on towards the target, steered by an observer in the nose of the launching aircraft operating a joystick controller linked to a radio transmitter. A flare in the tail of the weapon assisted tracking, and the observer's task was to superimpose the flare on the target and hold it there until the weapon impacted. To allow a number of aircraft to deliver simultaneous glider-bomb attacks without causing interference with the others' guidance signals, each Hs 293 was pre-set to use one of eighteen separate radio channels.

Early in 1942 a second type of radio-controlled guided missile began flight tests, the Ruhrstahl Fritz-X. This was an unpowered free-fall bomb weighing 3,100lb, with fixed cruciform wings mid-way along the body to give stability and movable control spoilers in the tail to allow it to be steered in flight. After release, the weapon could be steered on to the target using a similar radio guidance system to that fitted to the Hs 293. The Fritz-X was designed for use against armoured warships and, provided it was released from altitudes above 20,000 feet, it achieved an impact velocity that was sufficient to penetrate the deck armour of a battleship.

After a period of development and testing lasting over a year, the two new weapons entered series production in the spring of 1943. Two *Gruppen* of *Kampfgeschwader 100* were re-equipped with versions of the Dornier Do 217 specially modified to carry each of the missiles: *II Gruppe* received the E-5 variant able to carry one Hs 293 under each outer wing section; and *III Gruppe* re-equipped with the K-2 variant which could carry one Fritz-X bomb under each inner wing between the engine and the fuselage. The K-2 had a larger wing with a span 19ft greater than that of other variants of the bomber, to enable it to haul the Fritz-X bombs above 20,000ft so that they could achieve the required velocity during their fall to penetrate heavily armoured targets. Both *Gruppen* underwent a period of intensive training with their new aircraft, operating over the Baltic.

Early in July *III Gruppe* moved to its operational base at Marseilles/Istres in the south of France. The unit began flying combat missions soon afterwards, but initially the performance of the Fritz-X was disappointing. At dusk on the 21st, three aircraft attacked Allied ships in the port of Augusta in Sicily. No hits were scored. An attack on ships off Syracuse two days later was no more successful, nor were those against ships off Palermo and Syracuse on 1 and 10 August respectively. More seriously, the last two missions cost the *Gruppe* three Dorniers and their crews.

Meanwhile *II Gruppe* had moved to its operational base at Cognac in the west of France. The unit first went into action with its glider bombs on 25 August, when *Hauptmann* Heinz Molinnus led twelve Dorniers in an attack on a Royal Navy submarine-hunting group off the north-western tip of Spain. Each aircraft carried an Hs 293 only under the starboard wing, and a 66-gallon drop tank under the port wing. This attack was little more successful than the previous ones with the Fritz-X, and only the destroyer HMS *Landguard* suffered damage from a near-miss.

During a follow-up attack two days later, *II Gruppe* was much more successful. A force of eighteen Dorniers attacked another submarine-hunting group and this time a glider bomb scored a direct hit on the sloop HMS *Egret*. The explosion started a fierce fire that detonated the ammunition in the ship's after magazine. The ship broke up and sank with heavy loss of life. During the same action the destroyer HMCS *Athabaskan* suffered blast and splinter damage when a missile exploded in the water close to her, but again most of the glider bombs missed their targets.

It was clear that a disconcertingly large proportion of the missiles were failing to guide properly after launch, and a technical investi-

---

RUHRSTAHL FRITZ-X

Role: Air-launched guided missile, designed primarily for use against heavily armoured warships.
Powerplant: None.
Warhead: 3,100lb armour-piercing bomb containing 660lb of high explosive.
Performance: The weapon fell under gravity and its impact speed depended upon the altitude at which it was released. If released from 22,000ft it reached an impact speed of about 750mph.
Dimensions: Span (over cruciform fins) 4ft 3in; length 10ft 3¼in.
Method of guidance: Radio command-to-line-of sight.
Date first used in action: August 1943.

---

gation was ordered. This quickly unearthed evidence of sabotage to the aircraft. *Feldwebel* Fritz Trenkle, a technician working on the radio systems associated with the missile, explained how it was done:

> The command guidance signals from the aircraft transmitter were carried to the antenna via a coaxial cable, and somebody had cut the central conducting wire half-way along its length and then reassembled the cable. It was very clever, and obviously done by an expert. When we tested the transmitters on the ground with the aircraft engines stopped, the central conducting wire made a good enough contact and the signals were radiated properly. But when the engines were running the vibration caused the gap in the wire to open and close so that for long periods the guidance signals never reached the antenna. Once we had discovered the reason for the failure we checked all the Henschel 293-carrying aircraft and found that about half had been 'doctored' in this way.

The German security service carried out exhaustive inquiries in an effort to find the culprit, but without success. We shall never know how many Allied sailors owe their lives to the stealth and skill of the nameless saboteur. Once the reason for the failures was known, the faults were quickly cleared, and suddenly the new guided missiles began to demonstrate an awesome destructive power.

On 9 September the Italian government announced that it was making peace with the Allies. Under the terms of the armistice the Italian battle fleet was to sail to Malta to surrender. *Major* Bernhard Jope, the commander of *III Gruppe*, had been briefed on such a

---

DORNIER Do 217K-2

Role: Four-seat bomber and missile carrier.

Powerplant: Two BMW 801D 14-cylinder, air-cooled, radial engines each developing 1,700hp at take-off.

Armament: Normal operational bomb load about 4,900lb (up to 2,000lb more could be carried for short-range attacks); or one or (for short-range attacks) two Fritz-X guided bombs each weighing 3,100lb and carried on external racks on the wing between each engine and the fuselage. Defensive armament of one MG 151 15mm machine gun in top turret, two MG 131 13mm machine guns and three MG 81 paired 7.9mm guns on flexible mountings, firing forwards, rearwards and from either side of the cabin.

Performance: Maximum speed 313mph at 12,600ft; maximum cruising speed 280mph at 12,600ft; normal attack speed with Fritz-X missile, 290mph at 21,000ft; radius of action with one Fritz-X bomb with normal operational reserves, 820 miles.

Normal operational take-off weight: 36,625lb.

Dimensions: Span 80ft 4½in; length 55ft 9¼in; wing area 721 sq ft.

Date of first production Do 217K-2: Spring 1943.

possibility a week earlier (German cryptographers had been able to read the Italian naval cyphers for much of the war). Thus for several days, while Germany and Italy were nominally fighting on the same side, Jope had aircraft standing by at Istres ready to attack warships of his 'ally'.

As the armada of three battleships, six cruisers and eight destroyers neared the Strait of Bonifacio between Corsica and Sardinia, Jope at the head of twelve Dorniers caught up with the vessels. Each aircraft carried one Fritz-X and, concentrating on the battleships, the bombers ran in to deliver attacks from altitudes of around 22,000ft. Despite violent evasive action by the ships and a vigorous anti-aircraft defence, one of the missiles scored a direct hit on the Italian flagship, the modern battleship *Roma*. Impacting at a speed of about 750mph, the weapon struck amidships and punched clean through the hull before detonating immediately under the ship. The powerful explosion wrecked the steam turbines on the starboard side of the warship, causing her to lose speed. A few minutes later *Roma* took a second hit just forward of the bridge, which put the rest of her machinery out of action. The warship lost speed rapidly, with fierce fires raging inside the hull. Part of the blaze reached the forward magazine and touched off the ammunition stored there, causing a huge explosion. Her hull already weakened by the first bomb, the ship broke into two and sank with most of her crew. During the same action *Italia*, a sister-ship of *Roma* and formerly named *Littorio*, was hit by one Fritz-X. She took on 800 tons of water but was able to continue to Malta at a slightly reduced speed. The missiles also inflicted damage on two destroyers.

On the same day Allied troops landed at Salerno in southern Italy, and in the days that followed the concentration of shipping off the coast came under repeated attack from the missile-carrying Dorniers. On 11 September a Fritz-X hit the cruiser USS *Savannah* on the forward gun turret and detonated in the ammunition handling room, killing or injuring 270 of the crew. The force of the explosion punched a hole in the ship's bottom, opening a seam that allowed in sea water that extinguished the ship's boilers. Only prompt action by the damage control teams saved the ship. Two days later a similar attack on the Royal Navy cruiser HMS *Uganda* caused severe damage which required repairs lasting more than a year.

On 16 September the Royal Navy battleship *Warspite*, operating close inshore to provide bombardment support for the ground troops, came under attack from three Fritz-Xs. One struck amidships near the funnel and penetrated six decks to explode on, and blow a hole though, her double bottom. The other two bombs scored near-misses that tore gashes in her starboard side compartments. The ship lost steam and could not be steered, her radar and armament systems ceased to function and she took on 5,000 tons of water. The battleship was towed to Malta for repairs but was out of action for nearly a year.

II Gruppe was also in action during this period, but in several cases it is difficult to determine which type of guided missile caused the loss of a particular ship that was sunk. During the six weeks following the Salerno landings the missile-carrying Dorniers caused serious damage to four other warships and sank four small vessels. As the Allied bridgehead became established the fighter defences in the area stiffened markedly, and during the 42-day period the two *Gruppen* lost twelve crews in action. For the remainder of the year the missile-carrying Dorniers went into action from time to time against Allied convoys passing through the Mediterranean. In one such engagement, against a convoy off Oran on 11 November, sixteen Do 217s of *II Gruppe* took part in a co-ordinated attack together with forty He 111 and Ju 88 torpedo bombers from *Kampfgeschwader 26*. Three large freighters and a tanker were sunk, for the loss of one Dornier and six of the torpedo bombers.

With only incomplete information on the working of the German missiles, the US Naval Research Laboratory at Annapolis turned out a small batch of makeshift jamming transmitters. Two of these were installed in the destroyers USS *Davies* and *Jones*, which were sent to the Mediterranean.

In November 1943 a third *Luftwaffe* unit became operational with Hs 293s, *II Gruppe* of *Kampfgeschwader 40*, equipped with Heinkel He 177 heavy bombers. The first action by the *Gruppe* using glider bombs, on the 21st against a convoy passing to the west of the Bay of Biscay, resulted in one merchant ship being sunk and another damaged.

On the 26th the Heinkel 177s were in action again, when *Major* Rudolf Mons led twenty-one of them, each loaded with two glider bombs, to attack a convoy near Algiers. The destroyers *Davies* and *Jones* formed part of the convoy escort and they began jamming as soon as the missile guidance transmissions were picked up. As would later become clear, however, the jamming modulation was ineffective and it is doubtful whether it deflected any of the missiles. Glider bombs scored hits on the large troopship *Rohna* and she sank with heavy loss of life. Then Allied fighters arrived on the scene, and their vigorous counter-attack forced the Heinkels to abandon their missiles in flight or jettison those remaining. Six of the big bombers were shot down and others suffered damage. Rudolf Mons, one of the most successful German anti-shipping pilots, was among those killed.

The next major Allied landing operation in the Mediterranean was at Anzio, on 21 January 1944. The Dorniers of *II./KG 100* were again in action but by now the Allies had learned the crucial importance of providing strong fighter cover and establishing powerful anti-aircraft gun defences ashore as soon as possible after the landings. As a result there were few easy pickings for the missile-carrying bombers. During the 3½ weeks following the landings glider bombs sank the cruiser HMS *Spartan*, two destroyers and at least four transports, for a loss of seven crews from the *Gruppe*.

Representative of the attacks during this period by *II Gruppe* was that on the evening of 15 February. Nine of its Do 217s, each carrying two glider bombs, set out from Bergamo near Milan to attack shipping off the beach-head. It was almost dark when the bombers, flying singly, arrived at the target. *Oberfeldwebel* Paul Balke, flight engineer in one of the Dorniers, described the action from his viewpoint:

When we arrived the attack was already in progress. We did not see any other Dorniers, but from the ferocity of the flak it was obvious they were in the area. The ships put up a terrific barrage, as did the batteries on land; also some of the warships began laying smoke screens. We climbed to about 2,500 metres [8,000ft] and circled the target, then picked out a ship and headed towards it. *Hauptmann* Schacke, the observer, released the left bomb at a range of 7km [about 4 miles], which meant the missile had a flight time of about 50 seconds. After launch we could see the red flare in the tail of the missile clearly as Schacke guided it towards the ship. From where I saw it, the flare seemed almost stationary; each time it moved slightly off the target, Schacke guided it back on again. As the range increased the flare gradually became fainter and fainter, then there was a white explosion as it struck the ship almost amidships.

The Dornier turned and sped out of the defended area, then turned for a second attack and the crew lined up on another ship. Soon after the second glider bomb was launched, however, several of the ships concentrated their fire on the bomber. In the resultant profusion of converging tracer rounds Schacke lost sight of the missile's tracking flare and had to abandon it.

That evening glider bombs sank the 7,000-ton transport *Elihu Yale* and a tank landing craft. Despite the ferocity of the defences, all the Dorniers returned safely to Bergamo; usually they were not so fortunate.

By now the Allies had captured an intact example of the Hs 293 guidance receiver, and it quickly surrendered its secrets. Once its exact method of operation had been discovered, the Naval Research Laboratory developed the high-powered MAS jammer to counter the German guided missiles. The weakness of the missiles' guidance system was that it employed tone modulations on 1, 1.5, 8 and 12KHz to carry the appropriate up, down, left or right command to the weapon in flight. The MAS jammer concentrated all of its power on just one of the tone modulation frequencies, thus making the most efficient use of the available power. During tests against the captured receiver the MAS equipment demonstrated its effectiveness, and the jammer was ordered into production. Fifty of these transmitters were built and installed in escort vessels.

The *Luftwaffe* confidently expected that Hs 293 and Fritz-X missiles would be able to inflict severe losses when the Allies launched their long-awaited invasion of northern Europe. The specialized anti-shipping units were grouped together in *Fliegerkorps X*, based in the south of France under the command of *Generalleutnant* Alexander Holle. Previous attacks using the guided missiles had been made with forces of 25 aircraft or fewer. By careful husbanding of resources, by the end of May 1944 Holle had assembled a formidable force of missile carriers—four *Gruppen* with a total of 136 Heinkel He 177, Dornier Do 217 and Focke Wulf FW 200 aircraft. Supplementing these were four *Gruppen* with 136 Junkers Ju 88 torpedo bombers.

After dark on the evening of D-Day, 6 June 1944, *Fliegerkorps X* launched about forty missile-carriers and torpedo bombers to attack the concentrations of Allied shipping off the beach-head. When the bombers arrived in the invasion area they encountered a violent reception from the ships' gunners, while escort vessels added to their problems by laying smokescreens and radiating jamming on the missiles' radio-guidance channels.

Quite apart from these distractions, night attacks with the missiles proved much more difficult than by day. In the case of the Hs 293 there was a 2½-mile minimum launch range, and under operational conditions it was necessary to acquire the target initially at twice that distance to give time to align the aircraft on the target before missile launch. Unless there were clear moonlight conditions it was necessary to have other planes drop flares to illuminate the targets. With so many ships in the invasion area, it was difficult to co-ordinate the activities of the flare-droppers and the attack planes. As a result, the ships illuminated were often too far away from the missile-carriers; and by the time the latter had manoeuvred into position to deliver their attacks the flares had burned out. Because of the strength of the defences, the attacks had usually to be carried out in great haste, and with the combination of darkness, the jamming from the MAS transmitters and the smokescreens, the effectiveness of the glider bombs was greatly reduced.

A few aircraft attempted to deliver attacks with Fritz-X missiles but in order to score hits these weapons required the target to be visible from at least 15,000ft. Clear skies were thus an essential prerequisite for a successful attack, and such conditions were not often met over the English Channel.

In the weeks following D-Day, *Fliegerkorps X* smashed itself against the powerful defences protecting the concentrations of Allied shipping. There were few successes to show for its efforts: during the two weeks following the invasion only two Allied ships were sunk and seven damaged as a result of air attacks using conventional bombs or guided missiles.

Throughout most of their operational career the glider bombs and guided bombs had been restricted to attacks on naval targets, to reduce the likelihood of an example of either weapon falling into Allied hands (an embargo that continued long after the Allies had captured sufficient parts of both types of weapon to deduce their methods of operation). However, at the end of July 1944 US forces seized intact the bridges over the See and Sélune rivers at the southern end of the Cherbourg Peninsula, and several divisions streamed over the bridges before fanning out into the undefended countryside beyond. In the desperate attempts to slow the advance, the *Luftwaffe* received clearance to employ the Hs 293 against land targets for the first time. Do 217s of *III Gruppe* of *KG 100* attacked the bridges on the nights of 2, 4, 5 and 6 August, losing six aircraft and crews in the process. One of the bridges was hit, but the damage was not serious and after makeshift repairs by US Army engineers it continued in use.

Following the withdrawal of the remnants of *Fliegerkorps X* from France in August 1944, the Fritz-X was not used in action again and the Hs 293 saw little further use. The last recorded attack with the glider bombs was on 12 April 1945, when a dozen Do 217s flown by crews of *Kampfgeschwader 200* launched these weapons at bridges over the Oder river that had been captured intact by Soviet forces and over which forces were streaming westwards. Although the attackers claimed some hits, most of the bridges remained in use and the relentless Soviet advance continued.

The successes achieved by the Hs 293 and the Fritz-X fully justified the effort that was put into their development and production. Nearly all of those successes took place within a couple of months of the operational introduction of these weapons, however, and from then on their effectiveness deteriorated steadily. There were two reasons for this. First, it should always be borne in mind that an air-launched weapon is not one jot better than the ability of the carrying aircraft to bring it to within range of the target and, in the case of each of the two German weapons, to guide it throughout its flight until it impacted. Once the Allies knew of the existence of the new weapons, only rarely would a lucrative maritime targets appear within range of the German bombers without strong fighter protection. If the missile was in flight and the launching plane had to manoeuvre to avoid fighter attack, further guidance was made impossible. The second point is that if a new weapon is used in action, it is almost inevitable that examples will fall into enemy hands intact. When it does, the weaknesses of the system will be discovered and the enemy will exploit these in developing suitable countermeasures. In the case of the German guided missiles, the Achilles' heel was the tone-modulated guidance system that was vulnerable to the MAS jamming transmissions. Once the new jammer had been deployed aboard Allied naval escort vessels, the German weapons were able to achieve little.

6

HARD FIGHT TO 'THE BIG B'

The US bomber offensive against Germany sparked off some of the largest air actions in history. On 6 March 1944 the Eighth Air Force fought its hardest battle against the Luftwaffe, and suffered its heaviest losses, in the course of the first maximum-effort daylight attack to 'The Big B'—Berlin. This chapter tells the story of that engagement.

B Y THE BEGINNING of March 1944 the US Eighth Air Force considered itself sufficiently strong to take on the ultimate challenge—a daylight maximum-effort strike on Berlin, the most heavily defended target in Germany. After a couple of false starts, the attack took place on 6 March.

A total of 563 B-17 Flying Fortresses and 249 B-24 Liberators were assigned to the Berlin mission. The 1st Bomb Division, with 301 B-17s in five Wing formations, was to attack the VKF ball-bearing factory at Erkner, the third largest plant of its kind in Germany. The 2nd Bomb Division, with 249 B-24 Liberators in three Wing formations, was to bomb the Daimler-Benz works at Genshagen, then turning out more than a thousand aero engines per month. The 3rd Bomb Division, with 262 B-17s in six Wing formations, was to strike at the Bosch factory at Klein Machnow, which manufactured electrical equipment for aircraft and military vehicles.

For such a lengthy penetration of enemy airspace—800 miles from the Dutch coast to Berlin and back—much would depend on the ability of the escorting fighters to ward off the inevitable attacks by German fighters. Fifteen Groups of P-38 Lightnings, P-47 Thunderbolts and P-51 Mustangs of the Eighth Air Force, four Groups of Thunderbolts and Mustangs of the Ninth Air Force and three squadrons of RAF Mustangs—a total of 691 fighters—were to support the operation. After they had covered the bombers' initial penetration, the plan called for 130 Thunderbolts to return to the base, refuel and re-arm, then return to eastern Holland to cover the final part of the bombers' withdrawal.

Numerically, the escorting force was formidable, but two factors imposed limits on the number of escorts in position to protect the bombers if they came under attack from German fighters. The first constraint was the limited radius of action of the escorts: with drop tanks the fighters could penetrate deep into Germany, but only if they flew in a straight line. When accompanying bombers the escorts had to maintain fighting speed while matching their rate of advance with that of their slower charges. To do that they had to fly a zig-zag path, which added greatly to the distance flown. Furthermore, the escorts needed to retain a reserve of fuel in case they went into combat. These factors limited the time a fighter Group spent with the bombers over Germany to about half an hour, or 100 miles of the bombers' penetration. Then, it was hoped, another Group of fighters would relieve it. Thus the escort of a deep-penetration attack resembled a relay race, with some fighter units moving out to join the bombers, some with the bombers and others returning after completing their time with the bombers. At any one time there would be only about 140 Allied fighters flying with the bombers, less than one-sixth of the number of sorties they would fly that day.

The second serious constraint on the escorts was the great length of the bomber stream: 94 miles during that attack on Berlin. Had the 140 escorts been spread out evenly throughout that distance, there would have been just three fighters to cover every two miles of the bomber stream. Tactically that would have been a useless distribution. The solution was to concentrate about half the escorts around the one or two Combat Bomb Wings at the head of the bomber stream, for that part of the formation was the most likely to come under attack from German fighters. The remaining escorts were split into eight-plane flights that patrolled the flanks of the rest of the bomber stream. The arrangement meant that at any one time most of the bomber Wing formations had no escorts in a position to respond immediately if they came under attack from German fighters. Until help arrived, the bombers would have to rely on their own defensive firepower to hold off their attackers.

Starting at 7.50 a.m., the bombers of the Eighth Air Force began taking off from their bases in eastern England. Once airborne, they assembled into Group formations, then the Groups joined up to form Combat Bomb Wings. As the Bomb Wings crossed the coast of England at designated places and times, they slotted into position to form Divisions. At 10.53 a.m. the vanguard of leading Bomb Division, the 1st, crossed the Dutch coast a little over three hours after the first plane had taken off.

As they assembled into formation, the bombers were under the attentive gaze of the *Mammut* and *Wassermann* long-range early-warning radar stations in Holland and Belgium. Their reports were passed to the fighter-control bunkers from which the air defence of German homeland was managed. The action about to open would be controlled from Headquarters 3rd Fighter Division near Arnhem in Holland, that of 2nd Fighter Division at Stade near Hamburg and Headquarters 1st Fighter Division at Döberitz near Berlin.

With the three Bomb Divisions in line astern, the 1st and the 3rd with Flying Fortresses, then the 2nd with Liberators, the bombers thundered eastward over Holland at three miles per minute and at altitudes of about 20,000ft. The armada took half an hour to pass a given point on the ground, presenting an awesomely impressive spectacle of military might to those watching.

Some commentators have likened the US heavy bomber actions over Germany to those fought by the *Luftwaffe* over England during the Battle of Britain in the summer of 1940. Both led to large-scale daylight combats in which a numerically inferior defending force strove to protect its homeland against attacks by enemy bombers with strong fighter escorts. To be sure, there were similarities between the two campaigns, but the far greater distances to the targets in Germany meant that the defenders had time to assemble forces and deliver a more measured response than had been possible for their British counterparts. During the Battle of Britain the German raiding forces took half an hour to reach London, one of their more distant targets, from the south coast. In 1944, the US bombers often had to spend four times as long over hostile territory to reach their targets in Germany. In contrast to the hectic British fighter scrambles of 1940, the German fighter controllers usually had ample time to prepare their response and direct their fighters into position. Certainly, that would be the case on 6 March 1944.

BOEING B-17G FLYING FORTRESS

Role: Four-engine heavy bomber.
Powerplant: Four Wright R-1820 turbo-supercharged, 14-cylinder, air-cooled, radial engines each developing 1,200hp at take-off.
Armament: The bomb load carried depended upon the distance to be flown. During the attack on Berlin these aircraft carried twen 500lb high-explosive bombs. The defensive armament comprised two Browning .5in machine guns in powered turrets in the nose, above and below the fuselage and in the tail, and one on a hand-held mounting in each waist gun position and in the radio operator's firing position above the fuselage.
Performance: Typical formation cruising speed with bomb load, 180mph at 22,000ft; demonstrated operational radius of action during the Berlin mission, 550 miles (this included fuel consumed in assembly into formation, additional fuel consumed flying in formation, operational fuel reserves and the carriage of a 5,000lb bomb load released near the mid-point of the flight).
Normal operational take-off weight: 55,000lb.
Dimensions: Span 103ft 9½in; length 74ft 4in; wing area 1,420 sq ft.
Date of first production B-17G: September 1943.

MESSERSCHMITT Me 410A

Role: Two-seat bomber-destroyer.
Powerplant: Two Daimler Benz DB 603A 12-cylinder, inline engines each developing 1,750hp at take-off.
Armament: Offensive armament of six MG 151 20mm cannon mounted in the nose; four launchers for 21cm Wgr rockets under the outer wings. Defensive armament of two MG 131 13mm machine guns fitted in two remotely controlled barbettes firing rearwards and to either side.
Performance: Maximum speed 350mph at 22,000ft.
Normal operational take-off weight: 21,275lb.
Dimensions: Span 53ft 7¾in; length 40ft 11½in; wing area 390 sq ft.
Date of first production Me 410A: March 1943.

At airfields throughout Germany, Holland, Belgium and northern France, fighter units were brought through the different stages of alert until the pilots were at cockpit-readiness, awaiting the order to take off. As the raiding force headed due east across Holland, the German fighter controllers could see that it was probably heading for a target in northern Germany.

On this day the *Luftwaffe* had just over 900 serviceable fighters available for the defence of the Reich. Eighty were twin-engine Messerschmitt Bf 110s and Me 410s, specialized bomber destroyers armed with batteries of heavy cannon and launchers for 21cm rockets. There were nearly 600 Messerschmitt 109 and Focke Wulf 190 single-engine fighters. Backing these were more than 200 night fighters, Messerschmitt Bf 110s and Junkers Ju 88s, that could also be used by day.

Just as there were operational constraints to limit the proportion of the escorting fighters available to protect US bombers at any one time, so there were other constraints that limited the proportion of the defending fighter force that could be put into action against them. The tyranny of distance imposed its will on attacker and defender alike, and the defending fighters had to be disposed to protect targets in France, Holland and Belgium as well as almost the whole of Germany. To bring into action those units based far from the bombers' route, for example in eastern France or southern Germany, would require considerable prescience on the part of the fighter controllers and not a little luck. Moreover, although the bomber-destroyers had the range to reach any part of Germany, these large and unwieldy machines were liable to suffer heavy losses if they were caught by the escorts. Because of this, the twin-engine fighters were limited to engagements east of the line Bremen–Kassel–Frankfurt. The night fighters, slowed by the weight of their radar equipment and the drag from its complex aerial arrays, were suitable only for picking off stragglers.

FOCKE WULF FW 190A-8

Role: Single-seat general-purpose day fighter.
Powerplant: One BMW 801D-2 14-cylinder, air-cooled, radial engine developing 1,770hp at take-off.
Armament: Four MG 151 20mm cannon in the wings and two MG 131 13mm machine guns above the engine.
Performance: Maximum speed 402mph at 18,000ft; climb to 19,650ft, 9min 54sec.
Normal operational take-off weight: 9,660lb.
Dimensions: Span 34ft 5½in; length 29ft 4½in; wing area 197 sq ft.
Date of first production FW 190A-8: January 1944.

The risk that escorts would be accompanying the bombers, and the heavy defensive armament of the latter, presented further problems for the German fighter pilots. In order to overwhelm the escort, the defenders needed to deliver their attack *en masse*, and it took some time to assemble the units into one large formation. To reduce the effectiveness of the bombers' return fire, the fighters were to deliver their attack head-on, and that required careful direction from the ground controllers and skilful tactical handling from the formation leader.

At 11 a.m., seven minutes after the leading bombers crossed the Dutch coast, the first of the German fighter units began taking off in order to contest the incursion—107 Bf 109s and FW 190s drawn from *I* and *II Gruppen* of *Jagdgeschwader 1*; *I*, *II* and *III Gruppen* of *JG 11*; and *III Gruppe* of *JG 54*. Once airborne, the fighters assembled into *Gruppe* formations then climbed to altitude. As they did so they converged on the distinctive oblong outline of Lake Steinhuder near Hannover, the designated assembly point for the battle formation. Although this 'Big Wing' was twice as large as any that Douglas Bader had led during the Battle of Britain, it would not suffer the same shortcomings of the earlier tactic. For one thing the American bomber formations occupied a much larger volume of airspace than their *Luftwaffe* counterparts in 1940. Furthermore, the head-on attack offered time for only a short firing pass. These factors ensured that fighters would rarely get into each other's way during the actual attack.

Escorting the bombers during the initial part of their penetration into Germany were 140 Thunderbolts from the 56th, 78th and 353rd Fighter Groups. Although the escorts outnumbered the German fighters now preparing to engage the bombers, the arithmetic of the ensuing engagement was not on their side. The German fighter pilots would focus their attack on one Combat Bomb Wing formation, but until the last minute the escorts would be ignorant of where the blow would fall and so they had to divide their forces among several Wing formations.

Soon after the vanguard of the attacking force crossed the Dutch coast, the pathfinder B-17 at the head of the attack force suffered a radar failure. As a result, it flew a heading that took it a slightly south of the planned route. The rest of the bombers in the 1st Bomb Division followed it, as did those at the head of the 3rd Division. Before the leader had deviated 20 miles from the planned track the error was discovered, and the pathfinder edged on to a more northerly heading to regain the planned route. But by then the damage had been done. The 13th Wing, situated mid-way along the bomber stream, was a couple of minutes late at the Dutch coast and it had lost visual contact with the 4th Wing ahead of it. Ignorant of the deviation from the briefed route by the bombers ahead of it, the 13th adhered to the flight plan. The Wing would soon pay a terrible price for this accumulation of relatively minor errors.

At 11.55 *Hauptmann* Rolf Hermichen, at the head of the German battle formation, sighted the swarm of black specks in front of him. He had seen this many times before: it was a formation of US heavy bombers, a long way away. The specks appeared almost stationary in his windscreen—the two forces were closing almost exactly head-on. The German ground controllers had done their work well. From that distance Hermichen knew that any escorts with these bombers would be too small to see; he hoped there would be none, but he was enough of a realist not to depend on it.

It was sheer bad luck for the American bomber crews involved that the formation now under threat was not that leading the bomber stream, where more than half of the escorting Thunderbolts were concentrated: it was the 13th Wing, that heading the detached second half of the bomber stream, and it was almost devoid of such protection. Lieutenant Robert Johnson of the 56th Fighter Group was to one side of the Wing when he suddenly noticed Hermichen's force closing in fast:

> I was on the left side of the bombers and going 180 degrees to them when I noticed a large box of planes coming at us at the same level. There were about forty or fifty to a box, and I saw two boxes at our level and one box at 27,000 or 28,000 feet. I called in to watch them, and then that they were FW 190s. There were only eight of us . . .'

The Thunderbolts attempted to disrupt the attack but the German pilots simply ignored them as they streaked for the bombers. The opposing forces met at noon, 21,000ft above the small German town of Haselünne, close to the Dutch border.

A head-on attack on an bomber required a high degree of skill from the fighter pilot if it was to succeed. Closing at a rate of 200yds a second, there was time only for a brief half-second burst from 500yds before he had to ease up on the stick to avoid colliding with his target. For experienced pilots like *Hauptmann* Anton Hackl, the fighter ace leading the Focke Wulfs of *III Gruppe* of *Jagdgeschwader 11* that day, that was quite sufficient:

> One accurate half-second burst from head-on [on a four-engine bomber] and a kill was guaranteed. Guaranteed!

Feldwebel Friedrich Ungar of *Jagdgeschwader 54*, flying a Bf 109, saw his rounds exploding against the engine of one of the bombers and pieces flying off it:

> There was no time for jubilation. The next thing I was inside the enemy formation trying to get through without ramming anyone. Nobody fired at me then, they were too concerned about hitting each other. When we emerged from the formation things got really hot; we had the tail gunners of some thirty bombers letting fly at us with everything they had. Together with part of our *Gruppe* I pulled sharply to the left and high, out to one side. Glancing back I saw the Fortress I had hit tip up and go down to the right, smoking strongly.

Sergeant Van Pinner, a top-turret gunner with the 100th Bomb Group, recalled that he had far more targets than he could possibly fire at:

> There were fighters everywhere. They seemed to come past in fours. I would engage the first three but then the fourth would be on to me before I could get my guns on him. I knew our aircraft was being hit real bad—we lost the ball turret gunner early in the fight . . .

The initial head-on attack was over in much less than a minute and then, almost in slow motion, a succession of mortally wounded Fortresses began to slide out of formation. The 13th Wing comprised A and B formations flying almost in line abreast with a mile between them. The B formation comprised 38 Flying Fortresses from the 100th and 390th Bomb Groups and its Low Box, with sixteen B-17s at the start of the action, suffered the worst. All six bombers of its High Squadron were shot down, as were two of the six in its Lead Squadron and two of the four in its Low Squadron.

Lieutenant John Harrison of the 100th Bomb Group, Captain of one of the bombers, gazed in disbelief as planes began to go down around him:

> The engine of one Fort burst into flames and soon the entire ship was afire. Another was burning from waist to tail. It seemed [that] both the pilot and copilot of another ship had been killed. It started towards us out of control. I moved the squadron over. Still it came. Again we moved. This time the stricken Fortress stalled, went up on its tail, then slid down.

Following the initial firing pass, the German fighters split into twos and fours and curved around to deliver fresh attacks on the same formation. Some overtook the bombers and sped ahead of them preparatory to moving into position for further head-on attacks. Other German fighters attacked the bomber formation from behind, and yet others dived after damaged B-17s that had been forced to leave the formation and were trying to escape to the west.

Lieutenant Lowell Watts, Captain of a bomber in the next formation in the stream, was an unwilling spectator to the unequal battle:

> About two or three miles ahead of us was the 13th Combat Wing. Their formation had tightened up since I last looked at it. Little dots that were German fighters were diving into those formation, circling, and attacking again. Out of one High Squadron a B-17 slowly climbed away from its formation, the entire right wing a mass of flames. I looked again a second later. There was a flash—then nothing but little specks drifting, tumbling down. Seconds later another bomber tipped up on a wing, rolled over and dove straight for the ground. Little white puffs of parachutes began to float beneath us, then fall behind as we flew toward our target.

From the moment the German fighters had first been sighted, the 13th Wing put out frantic radio calls to summon assistance from the escorts. Colonel Hub Zemke, commander of the 56th Fighter Group and heading an eight-plane flight of Thunderbolts, arrived at the beleaguered unit just as *Oberleutnant* Wolfgang Kretschmer of *JG 1* was lining up for another firing pass. Zemke spotted the lone Focke Wulf below him and ordered one section of four aircraft to remain at high altitude to cover him, while he led his section down to attack.

Before opening fire at the bomber he had selected as a target, Kretschmer glanced over his own tail to check that the sky was clear. It was not. The German pilot was horrified to see Zemke's Thunderbolt closing in rapidly on him, followed by three others. Kretschmer hauled the Focke Wulf into a tight turn to the left get out of the way, but it was too late. By then Zemke was in a firing position and .5in rounds from his accurate burst thudded into the wings and fuselage of the German fighter. As Zemke pulled up to regain altitude he glanced back and saw the enemy fighter falling out of the sky enveloped in flames. Kretschmer extricated himself from the cockpit of his blazing aircraft and jumped clear. He landed by parachute with moderate burns to his hands and face and splinters embedded in his thigh.

The main part of the initial action lasted about ten minutes. Then, as the German fighters exhausted their ammunition, they dived away from the fight, trying to avoid the rampaging escorts. However, even as the initial action petered to its close, a second German battle formation was already moving into position to engage the raiders. In their bunker at Döberitz the controllers of the 1st Fighter Division had assembled every available fighter in that part of Germany. The core of the battle formation comprised the bomber-destroyers, 42 Messerschmitt Bf 110s and Me 410s from *II* and *III Gruppen* of *Zerstörergeschwader 26* and *I* and *II* of *ZG 76*. Providing cover for these, though they were also expected to engage the bombers, were 70 Bf 109s and FW 190s from *I*, *II* and *IV Gruppen* of *Jagdgeschwader 3*, *I* of *JG 302* and *Sturmstaffel 1*. Leading the formation was *Major* Hans Kogler, the commander of *III./ZG 26*, flying a Bf 110.

Again, a large force of German fighters charged head-on into a pair of formations of Fortresses flying in line abreast: the 1st Combat Wing comprised 51 aircraft drawn from the 91st and 381st Bomb Groups; and the 94th Combat Wing, with 61 Flying Fortresses from the 401st and 457th Bomb Groups, flew a couple of miles to the right of it. But these two Wings were in the vanguard of the bomber stream and were protected by a large proportion of the available escorts—80 Mustangs from the 4th and 354th Fighter Groups. This time the escorts were in the right place, at the right time and in sufficient numbers to blunt the German attack. Lieutenant Nicholas Megura of the 4th Fighter Group described the approach of the defending formation:

> Twelve-plus smoke-trails were seen coming from twelve o'clock and high, thirty miles ahead. 'Upper' [the Group leader] positioned the Group up-sun, below condensation height, and waited. Trails finally positioned themselves at nine o'clock to bombers and started to close. Six thousand feet below the trails were twenty-plus single-engine fighters line abreast, sweeping area for twenty-plus twin-engine rocket-carrying aircraft. 'Upper' led Group head-on into front wave of enemy aircraft.

The Mustangs' spoiling tactics forced several German fighters to abandon their attack, but others continued doggedly on to launch their hefty 21cm-calibre rockets head-on into the bomber formations.

Accidentally or deliberately, an Me 410 collided with, or rammed head-on into, a Flying Fortress of the 457th Bomb Group and tore away a large section of the bomber's tail. The stricken bomber, which had been flying on the right side of the High Squadron, went out of control and entered a steep, diving turn to the left. After narrowly missing several bombers in the formation, it smashed into the aircraft on the far left of the Low Squadron. Only one man survived from the three crews involved in the incident, the tail gunner from the last aircraft to be struck. As the bomber-destroyers emerged from the rear of the bomber formation, other Mustangs pounced on them and the nimble single-seaters did great execution; fourteen of the twin-engine fighters were shot down.

Hard on the heels of the heavy fighters came the main body of the attack formation—70 single-seat Messerschmitts and Focke Wulfs. One of the attacking pilots was *Leutnant* Hans Iffland of *Jagdgeschwader 3*, flying a Messerschmitt Bf 109:

> During the firing run everything happened very quickly, with the closing speed of about 800 kilometres per hour [500mph]. After firing my short burst at one of the B-17s I pulled up over it; I had attacked from slightly above, allowing a slight deflection angle and aiming at the nose. I saw my rounds exploding around the wing root and tracers rounds from the bombers flashing past me. As I pulled up over the bomber I dropped my left wing, to see the results of my attack and also to give the smallest possible target at which their gunners could aim. Pieces broke off the bomber and it began to slide out of the formation.

The action around the leading formations lasted little over ten minutes and, thanks to the efforts of the Mustangs, the bombers' losses were much lighter than during the earlier attack: seven had been either destroyed or damaged so seriously that they had been forced out of formation and finished off as stragglers.

Now, shortly after 1 p.m., the raiding units were moving into position to begin their bomb runs. Defending Berlin was the 1st Flak Division, commanded by *Generalmajor* Max Schaller and comprising the 22nd, 53rd, 126th and 172nd Flak Regiments with more than 400 8.8cm, 10.5cm, and 12.8cm guns. German fighter pilots were unwilling to pursue the raiders into the inferno of flak that would soon be put up over the capital, and they broke off the action.

As they approached their targets, the bombers split into their three divisions and lined up for their bombing runs. The first to feel the gunners' wrath were the Flying Fortresses of the 1st Bomb Division. Captain Ed Curry, a bombardier with the 401st Bomb Group, never forgot that cannonade:

I'd been to Oschersleben and the Ruhr, but I'd never seen flak as heavy as that they had over Berlin. It wasn't just the odd black puff, it was completely dense; not just at one altitude, but high and low. There was a saying that you see the smoke only after the explosion; but that day we actually saw the red of the explosions. One shell burst near us, and we had chunks of shell tear through the radio room and the bomb bay.

Now, however, the weather protected the primary targets more effectively than the German defences ever could. At first it seemed that the bombardiers would be able to make visual bomb runs on the targets through breaks in the clouds, but at the critical moment the aiming points were obscured. None of the planes hit the 1st Division's primary target at Erkner, and the attackers released their bombs on the Köpenick and Weissensee districts of the city.

It was a similar story for the 3rd Bomb Division, whose Flying Fortresses missed the primary target at Klein Machnow and bombed the Steglitz and Zehlendorf districts instead. Lowell Watts was on his bomb run when the gunners zeroed in on his formation:

They didn't start out with wild shots and work in closer. The first salvo they sent up was right on us. We could hear the metal of our plane rend and tear as each volley exploded. The hits weren't direct. They were just far enough away so they didn't take off a wing, the tail or blow the plane up; they would just tear a ship half apart without completely knocking it out. Big ragged holes appeared in the wings and the fuselage. Kennedy, the co-pilot, was watching nothing but the instruments, waiting for the tell-tale indication of a damaged or ruined engine. But they kept up their steady roar, even as the ship rocked from the nearness of the flak bursts . . . The flak was coming up as close as ever, increasing in intensity. Above and to the right of us a string of bombs trailed out from our lead ship. Simultaneously our ship jumped upwards, relieved of its explosive load as the call 'Bombs away!' came over the interphone. Our left wing ship, one engine feathered, dropped behind the formation. That left only four of us in the Low Squadron. A few minutes later the flak stopped. We had come through it and all four engines were still purring away.

Only a few Liberators of the 2nd Bomb Division, the last to attack, put their bombs on their primary target, the Daimler-Benz aero-engine works at Genshagen; the rest of the attack also fell on secondary targets in and around the capital.

The vicious bombardment knocked down only four bombers, but it caused sufficient damage to several others to force them to leave formation. Moreover, nearly half the bombers that reached Berlin collected flak damage of some sort.

As the bombers emerged from the flak zones, a few German single-seat fighters tried to a press home attacks, while fourteen Messerschmitt 110 night fighters from *Nachtjagdgeschwader 5* closed in to finish off stragglers. The escorts quickly took charge of the situation, however. The defending single-seaters were chased away and the night fighters, too slow to escape from their pursuers, lost ten of their number within the space of a few minutes. For the time being the German fighters had spent their force, and during the next half-hour there was a lull in the fighting.

Relieved by fresh squadrons of Thunderbolts, the Mustangs peeled away from the bombers and headed for home. As they were running out past Bremen, a section of Mustangs of the 357th Fighter Group came upon a lone Bf 109 and Lieutenants Howell and Carder shot it down. *Oberleutnant* Gerhard Loos of *Jagdgeschwader 54*, a leading *Experte* credited with 92 victories, was killed.

The air action around the bombers resumed at 2.40 p.m., with attacks by Bf 109s and FW 190s that had landed to refuel and re-arm after taking part in the noon action near Haselünne. Other fighters came from units based in France and Belgium that had missed the raiders on their way in.

The formation hardest hit during this engagement was the 45th Combat Bomb Wing. Once again Lowell Watts of the 388th Bomb Group takes up the story:

> The interphone snapped to life: 'Focke Wulfs at 3 o'clock level!' Yes, there they were—what seemed at a hurried count to be about 30 fighters flying along just out of range beside us. They pulled ahead of us, turned across our flight path and attacked from ahead and slightly below us. Turrets swung forward throughout the formation and began spitting out their .50-calibre challenge. Some Focke Wulfs pulled above us and hit us from behind while most dived in from the front, coming in from 11 to 1 o'clock to level, so close that only every second or third plane could be sighted on by the gunners. Still they came, rolling, firing and diving away, then attacking again.

He watched two bombers fall out of the formation, then his own aircraft came under attack:

> Brassfield called from the tail position, 'I've got one, I've got one!' Then, almost with the same breath, 'I've been hit!' No sooner had the interphone cleared from that message when an even more ominous one cracked into the headsets: 'We're on fire!' Looking forwards I saw a Focke Wulf coming at us from dead level at 12 o'clock. The fire from our top and chin turrets shook the B-17. At the same instant his wings lit up with

fire from his guns. The 20mm rounds crashed through our nose and exploded beneath my feet amongst the oxygen tanks. At the same time they slashed through some of the gasoline cross-feed lines. The flames which started here, fed by the pure oxygen and the gasoline, almost exploded through the front of the ship. The companionway to the nose, the cockpit and the bomb bays was a solid mass of flame.

Watts struggled to hold the bomber level while his crew abandoned the machine. The flames prevented him from seeing ahead and he could not know that his aircraft was edging ever closer to another formation. With a crash of tortured metal the bombers smashed together, then broke apart and, shedding pieces, they began their long fall to earth.

Unaware that there had been a collision, Watts knew that the bomber was no longer under his control. Moreover, seemingly for no good reason, almost the whole of the cabin roof above his head had suddenly vanished:

It was a wild ride from that point. I could tell we had rolled upside down. My safety belt had been unbuckled. I fell away from the seat, but held myself in with the grasp I had on the control wheel. After a few weird sensations I was pinned to the seat, unable to move or even raise my hand to pull off the throttles or try to cut the gas to the inboard engines. My left foot had fallen off the rudder bars while we were on our back. I couldn't even slide it across the floor to get it back on the pedal. Flames now swept past my face, between my legs and past my arms as though sucked by a giant vacuum. Unable to see, I could tell only that we were spinning and diving at a terrific rate. That wild eerie ride down the corridors of the sky in a flaming bomber still haunts my memory. But it wasn't just the terror of death, it was the unending confusion and pain of a hopeless fight and the worry for the nine other men that were my responsibility. Contrary to the usual stories, my past life failed to flash in review through my mind. I was too busy fighting to keep that life.

The next thing Watts knew, he was hurtling through space and well clear of the blazing bomber. He opened his parachute and landed safely, as did five others from his crew. Six men from the other bomber involved in the collision also parachuted to safety.

During this sharp engagement the 388th Bomb Group lost a total of seven aircraft. The losses were not all on one side, however. *Hauptmann* Hugo Frey of *Jagdgeschwader 11*, an ace pilot credited with the destruction of 26 US heavy bombers, was killed when his FW 190 was shot down by return fire from the bombers.

On 6 March a total of 812 Flying Fortresses and Liberators had set out from England, 672 of which had attacked primary or secondary targets in the Berlin area. Sixty-nine B-17s and B-24s failed to return to England, including four damaged machines that came down in Sweden. Sixty bombers returned with severe damage and 336 had lesser amounts of damage. Of the 691 fighters taking part in the operation, eleven were destroyed and eight returned with severe damage. In a concentrated fighter-versus-bomber action of this type, the heavy losses were confined to a few unfortunate units; as luck would have it, all of them flew Flying Fortresses. The hardest hit, the 100th Bomb Group, lost fifteen of the 36 aircraft that had crossed the Dutch coast; most of these fell during the initial clash near Haselünne. The 95th Bomb Group, which with the 100th made up the 13B Wing formation, lost eight bombers. The 91st Bomb Group lost six during the attack in front of Berlin, and the 388th Bomb Group lost seven planes during the sharp action over western Germany on the way home.

The losses suffered by these four Groups accounted for more than half the heavy bombers that failed to return. By its nature, an account such as this will tend to concentrate on the areas of heaviest fighting and the units that took the heaviest losses. To put things into perspective, it should be pointed out that the remaining 33 bomber losses were shared almost evenly between nineteen Bomb Groups; and the remaining six Bomb Groups flew the mission without suffering a single loss between them.

That day the *Luftwaffe* flew 528 fighter sorties, of which 369 probably made contact with the raiders. Sixty-two German fighters, 16 per cent of those that made contact, were destroyed and thirteen damaged. The twin-engine fighter units took the heaviest losses. *Nachtjagdgeschwader 5* lost ten of the fourteen Messerschmitt Bf 110 night fighters sent up, while *Zerstörergeschwader 26* lost eleven of its eighteen Messerschmitt Bf 110 and Me 410 bomber-destroyers. Altogether the *Luftwaffe* lost 44 air crew killed, including two leading aces, and 23 wounded. The aircraft losses could be made good, but the experienced fighter pilots were irreplaceable.

With the attack on Berlin on 6 March 1944 the Eighth Air Force bombing campaign passed another important milestone. Thenceforth no target in Germany, no matter how far from England nor how strong its defences, was immune from daylight precision attack.

7

THE GREAT MARIANAS TURKEY SHOOT

The Battle of the Philippine Sea in June 1944, involving fifteen US and nine Japanese aircraft carriers with full supporting forces, was the largest carrier-versus-carrier action of all time . . .

O N 6 JUNE 1944, as Allied troops fought to secure a foothold on the Normandy beaches, half a world away another large invasion force was making its opening moves. Comprising more than 600 ships bearing 127,000 assault troops, its objective was the Pacific island of Saipan in the strategically placed Marianas group.

The cutting edge of the US Pacific Fleet was Vice-Admiral Marc Mitscher's fast carrier striking force, Task Force 58. This comprised seven large aircraft carriers and eight smaller ones, with 467 Hellcat fighters. In addition there were 425 attack planes—Dauntless and Helldiver dive-bombers and Avenger multi-purpose bombers that could attack with bombs or torpedoes. The US carrier air groups were highly trained and proficient and most of their crews had several months' combat experience. They functioned as well-knit teams, confident in their equipment and in their ability to defeat their opponents.

In addition to its front-line force, Task Force 58 was well provided with reserves for the forthcoming battle. Since the action was planned to take place about a thousand miles away from the nearest US support base, at Eniwetok Atoll, the size of the force deployed against the enemy was limited to that which the available fleet logistics train could support. Because of this, no fewer than five escort carriers with a total of 135 aircraft were held in reserve off Eniwetok, ready to move into the combat zone to replace any carrier put out of action; and some 100 combat aircraft of all types, with crews, were at sea aboard the replenishment carriers sailing with the logistic support ships, available to replace losses. Operating separately from the fast carrier force were eight more escort carriers, with 114 Wildcat fighters and 82 Avenger bombers, earmarked to protect the transports and landing craft and provide air support for the troops as they went into action ashore.

Since its disastrous defeat during the Battle of Midway in June 1942, when the Japanese Navy lost four large carriers together with most of their air groups in a single day, that service had spared no effort to rebuild its aircraft carrier fleet. By the spring of 1944 it was once again a force to be reckoned with, with three large and six smaller carriers operating a total of 430 combat aircraft. In addition there were nearly 500 combat aircraft of all types based at airfields in the Marianas, and a further 100 in the Caroline Islands within range to intervene in the action about to begin.

On 11 June Task Force 58 arrived within attack range of the Marianas and began a programme of offensive strikes aimed at establishing air superiority over Guam, Saipan and Tinian. Two days later a force of battleships and their attendant cruisers and destroyers arrived off Saipan, to bombard coastal positions and provide cover for the minesweepers moving inshore to check that the approaches to the invasion beaches were clear of mines and other obstacles.

The importance of Saipan, and the deterioration to Japan's war situation if that island were lost, was all too clear to the Japanese High Command. The island lay within 1,500 miles of Japan itself, and therefore within B-29 heavy bomber range of the home islands. If the Marianas fell, Tokyo and every other Japanese city faced the menace of sustained aerial bombardment. The threat could not be ignored, and the Japanese Navy was prepared to stake everything it had to deliver an effective counterstroke.

On 13 June the Japanese carrier force, commanded by Vice-Admiral Jisaburo Ozawa, set sail from Tawi Tawi in Borneo and headed for the Marianas. One the way other groups of warships joined the force, and the Japanese Commander-in-Chief, Admiral Soemu Toyoda, repeated to each ship the famous signal that Admiral Togo had sent before the decisive Battle of Tsushima that had seen the defeat of the Imperial Russian Navy 38 years earlier: 'The fate of the Empire rests on this one battle. Every man is expected to do his utmost.'

To win the forthcoming engagement the Japanese sailors and airmen would certainly need courage and resourcefulness of a very high order. Their carrier planes were outnumbered more than two to one by those of Task Force 58 alone, and that was only one of their predicaments. The most serious problem was the general lack of combat experience, compounded in many cases by the incomplete operational training, of the Japanese air crews. The hastily reassembled Japanese carrier force had reached its present size only by the careful husbanding of resources and by keeping most of the flying units out of action. The air crews aboard the three larger carriers had six months' operational training and were considered proficient in their respective roles (though their level of training was considerably less than that received by their US Navy counterparts). Aboard the six

YOKOSUKA D4Y1 ('JUDY')

Role: Two-seat, carrier-borne dive-bomber/reconnaissance aircraft.
Powerplant: One Aichi Atsuta liquid-cooled, 12-cylinder, inline engine developing 1,200hp for take-off.
Armament: One 1,100lb bomb carried in an internal bomb bay; two sychronized Type 97 7.7mm machine guns firing through the propeller disc and one hand-held Type 1 7.92mm machine gun firing rearwards.
Performance: Maximum speed 332mph at 17,225ft.
Normal operational take-off weight: 8,276lb.
Dimensions: Span 37ft 8¾in; length 33ft 6½in; wing area 254 sq ft.
Date of first production D4Y1: October 1942.

smaller Japanese carriers the average level of training was much lower and many of the crews had not reached even the minimum Japanese operational training standard.

The unevenness in the quality of the Japanese air crews was matched by that of the aircraft they few. The best Japanese attack planes, the B6N ('Jill') and the D4Y1 ('Judy') that equipped the torpedo and dive-bomber units respectively on the three larger Japanese carriers were first-rate fighting aircraft with a good turn of speed and a radius of action superior to that of their American counterparts. On the six smaller Japanese carriers it was a different matter. Their decks were too small to operate the high performance 'Jills' and 'Judys', and their attack units were equipped with the obsolete B5N ('Kate') torpedo bombers and D3A1 ('Val') dive-bombers that dated from the beginning of the Pacific war. These small carriers also operated units equipped with the elderly A6M2 Model 21 Zero fighter modified as a dive-bomber with a rack to carry a single 550lb bomb under the fuselage. The modified Zeros lacking dive brakes, however, and they were limited to medium- or shallow-dive attacks that were less accurate against moving ships than steep-dive attacks.

NAKAJIMA B6N2 ('JILL')

Role: Three-seat, carrier-borne torpedo bomber.
Powerplant: One Mitsubishi Kasei 25 14-cylinder, air-cooled, radial engine developing 1,825hp at take-off.
Armament: One 1,760lb torpedo, or an equivalent weight of bombs; one sychronized Type 97 7.7mm machine gun firing forwards through the propeller disc, one firing rearwards from the cabin and one firing rearwards and below the aircraft from the ventral hatch.
Performance: Maximum speed 318mph at 19,500ft.
Normal operational take-off weight: 12,070lb.
Dimensions: Span 49ft; length 36ft 1in; wing area 395 sq ft.
Date of first production B6N2: October 1943.

The most serious equipment weakness faced by the Japanese carrier air groups was the relative obsolescence of their standard fleet air defence fighter type, the Mitsubishi A6M5 Type 52 (Allied code-name 'Zeke 52'). Developed from the earlier Zero, it was only slightly more effective, with a maximum speed increased by a paltry 20mph. Compared with the standard US Navy carrier fighter type, the Grumman F6F Hellcat, the Type 52 was inferior in almost every aspect of performance that mattered in air combat. Flights trials in which a captured A6M5 Model 52 was pitted in mock combats against Hellcats revealed huge discrepancies in the performance of the two aircraft. The report on the trials stated:

In maximum speed, the F6F ranged from 25 mph faster at sea level to 75 mph faster at 25,000 feet. In the climb the Model 52 was superior below 14,000 feet, at altitudes above that the F6F was superior. Below 230 mph the rate of roll of the two fighters was similar, above that speed the F6F-5 was much the better. Below 200 mph the Model 52 was far more manoeuvrable than the F6F, while above 230 mph the F6F was the more manoeuvrable.

Hellcat pilots engaging the Zeke 52 were urged to exploit to the full their superior speed and altitude performance. On no account were they to allow themselves to be drawn into dogfights with the more nimble but slower Japanese fighter. An effective tactic, when the Japanese fighters were sighted, was to increase speed to 300mph then enter a zoom climb, converting speed into altitude to get well above and out of reach of their opponents. Once there the American pilots had the initiative and could dictate the terms of the engagement, delivering high-speed slashing attacks on the Zekes followed by a zoom climb back to altitude to allow no opportunity for retaliation. Under such conditions the Japanese fighter's superior low-speed manoeuvrability would be an unusable asset.

On 15 June US Marines stormed ashore on Saipan, and by nightfall they had established two beach-heads. This development gave added urgency to the Japanese fleet now approaching the island. During the initial phase of the battle Admiral Toyoda enjoyed three important advantages over the US fleet commander, Admiral Raymond Spruance. The first advantage was freedom of action: Toyoda could deploy his carriers as the tactical situation demanded, whereas Spruance had to keep his force of fast carriers close to Saipan to protect the landing operation and the vulnerable transport ships off the coast. Toyoda's second advantage was that his long-range aerial reconnaissance was much the more effective, and planes flying from the Marianas were already shadowing the various groups of US ships. For their part the Japanese carriers were still beyond the 350-mile radius of action of the US carrier reconnaissance planes, so Spruance had to rely on the fleeting and often misleading sighting reports from submarines for information on the whereabouts of his opponents. The third great advantage enjoyed by the Japanese commander was that the attack radius of action of his carrier planes, 300 miles, was a full 100 miles greater than that of the US planes that were burdened with greater armour protection and self-sealing fuel tanks.

Toyoda had regular reports on the movements and approximate compositions of the US Navy task groups, while Spruance had only a vague idea of the location of the Japanese carriers. So long as that situation continued, it was inevitable that the Japanese would be the first to launch an air strike and that the US ships would have to brace themselves to fight a defensive battle.

The first attacks on the US ships came not from the Japanese carrier planes, nor from those based in the Marianas, but from aircraft based at Yap in the Caroline Islands. In the afternoon of the 17th, a force of B5N 'Kate' torpedo bombers sank a Landing Craft (Infantry). That evening seventeen D4Y 'Judy' dive-bombers and two P1Y1 'Frances' torpedo bombers with a strong fighter escort attacked the transports and escort carriers off the beach-head and damaged one of the small carriers so seriously that she had to be withdrawn from the operation. In further air attacks the next day a fleet oiler was damaged.

Before dawn on the 19th Admiral Ozawa's carriers reached a position from which they could launch a massive air strike against the Task Force 58. Japanese reconnaissance planes increased their shadowing activity around the US task groups, an essential operation before the attack but also a costly one: of about 60 carrier and land-based aircraft assigned to the task, no fewer than 35 were shot down by the Hellcats.

Early that morning the Japanese carriers turned into the wind and began launching their planes. The operation did not take place unhindered, however, for the submarine USS *Albacore* was able to sneak past the escorting destroyers and loose off a spread of torpedoes at the large Japanese carrier *Taiho*. One of the weapons struck the warship, but it caused only minor damage to her hull and after a short delay *Taiho* resumed launching her planes.

The first of the incoming attack forces, comprising 43 Zero dive-bombers and seven B6N ('Jill') torpedo bombers with an escort of fourteen 'Zeke 52s', was detected on radar while still 150 miles from the US carriers. Any possible doubts regarding the force's intentions were dispelled soon afterwards when the Japanese air group commander delivered a radio briefing to his crews that was overheard by a monitor on board one of the US warships.

The US carriers immediately began to clear their decks for action. The Hellcats assembled into flights and climbed to the west to meet the threat while the attack planes headed east to keep out of harm's way until it was safe to return. When the Japanese force was detected there were 59 Hellcats airborne on combat air patrol to defend the task force, and now they were joined by a further 140 fighters.

The cloud-free skies and the ample flow of radar information on the raiders' approach provided ideal conditions for the defending fighters, and the latter made the most of their advantage. The first to go into action were eight Hellcats the USS *Essex*, which intercepted the Japanese force when it was still 70 miles from the American carriers. With a 6,000ft altitude advantage, Commander C. Brewster led a diving attack on the Zero dive-bombers that accounted for two of them, then the escorting 'Zekes' intervened and a wild mêlée developed. Soon afterwards twenty Hellcats from *Hornet* and *Princeton* pitched into the fight, followed by a further fourteen from *Cowpens* and *Monterey*. Other fighters, having recently taken off from their carriers, joined the action as the Japanese force closed on the US carriers. Subjected to a series of savage onslaughts from above, the raiding force took heavy losses. Some of the Japanese attack formations lost cohesion but their pilots refused to countenance retreat and pressed on grimly towards their targets. Despite the heavy odds against them, about twenty Japanese attack planes broke through the fighter screens and lined up to deliver attacks on the US warships.

It requires a large amount of training and practice to deliver an accurate dive-bombing or torpedo attack on a warship manoeuvring in open water, and the abilities of the Japanese crews did not match the iron determination they had displayed in getting through to their targets. Their sole hit was a bomb that struck the battleship *South Dakota*, causing localized damage and some fifty casualties, but did not prevent the warship continuing in action. Forty-two of the 64 attacking planes were shot down, for the loss of three Hellcats.

Hard on the heels of the first Japanese striking force came the second. Launched from the three largest Japanese aircraft carriers, this comprised 53 D4Y1 ('Judy') dive-bombers and 27 B6N ('Jill') torpedo bombers with an escort of 48 'Zeke 52' fighters. These were the most modern of the Japanese carrier attack aircraft and they were flown by the best-trained of the available crews. Again the striking force was detected on radar when it was more than 100 miles from the American carriers, and again a monitor eavesdropped on the airborne radio briefing given by the Japanese attack commander. Since the earlier launch there had been time to range more Hellcats on the decks of the US carriers. In addition to the fighters sent aloft to engage the first raid, many of which had fuel and ammunition to continue the fight, more than 160 fresh fighters were scrambled to buttress the defence.

The second action was a near-repeat of the first. Once more the raiders suffered heavy losses on the way in, but again about twenty broke through the defences. And once again the resultant attack highlighted the Japanese crews' poor training and lack of combat experience. The aircraft carriers *Wasp* and *Bunker Hill* suffered minor damage and some casualties from bombs exploding nearby, while other ships were narrowly missed by torpedoes. A 'Jill' crashed into the battleship *Indiana*, but the latter's armoured belt took the force of the explosion and the warship continued in operation. As before, the defending fighters did great execution, and of the 128 planes in the attack wave, 94 were shot down for a loss of four Hellcats.

Owing to an inaccurate report on the position of the target, most of the 49 aircraft in the third Japanese attack wave failed to reach the US task force. Only the escorting force of sixteen 'Zeke 52s' went into action, and the Hellcats shot down seven of the Japanese fighters for the loss of one of their own.

The fourth attack wave, comprising 46 'Judy', 'Val' and Zero dive-bombers, six 'Jill' torpedo bombers and an escort of thirty 'Zeke 52s', did no better than the first two. A few Japanese planes did break through the fighter screens and to deliver attacks to the US carriers, but none of them scored hits. Thirty-eight of the planes in this force were shot down, either in the vicinity of the US carriers or when they were caught by Hellcats as they were on the landing approach to Orote airfield on Guam.

During the series of air combats, four Japanese carrier-borne attack forces with a total of 323 aircraft had been hurled into action against Task Force 58. Of those aircraft, 181 were shot down—a crushing 60 per cent of the raiding force. Seven Hellcats had been lost in the swirling air combats, but none of the ships had suffered serious damage and all were able to continue operating.

Meanwhile the US submarines continued to nip at the heels of the Japanese carrier groups far to the west. Shortly after the survivors from the four attack waves returned to their ships, the USS *Cavalla* scored three torpedo hits on the large carrier *Shokaku*. In spite of desperate efforts by the latter's crew to contain the fires, these were soon blazing out of control and, following a large explosion, the vessel went to the bottom. Soon afterwards *Taiho* followed her; although the single torpedo hit earlier in the day had caused relatively little structural damage, fumes from a ruptured aviation fuel tank had leaked into the hull. A chance spark ignited these, leading to a series of explosions of increasing severity until finally the carrier sank. Twenty-two planes went down with the two Japanese carriers.

By dawn on 20 June the Japanese carrier air groups were reduced to a shadow of their former strength: they were left with just 68 'Zeke 52s' and Zeros, three D4Y 'Judy' dive-bombers and 29 B5N 'Kate' and B6N 'Jill' torpedo bombers. Ozawa had no alternative but to order his seven surviving carriers to withdraw to the west. On Guam the Japanese air strength was almost exhausted too, following the loss of 110 planes in the air and on the ground and many more damaged during the US attacks on airfields.

Despite the best efforts of the US carrier reconnaissance planes, it was mid-afternoon on the 20th before Admiral Mitscher received the first clear reports on the location of the Japanese carrier groups. At last, Task Force 58 could deliver its counterstroke. The striking force comprised 77 Helldiver and Dauntless dive-bombers, 54 Avenger torpedo bombers and an escort of 85 Hellcats. The attack was to be mounted at maximum range, and soon after the force set out there was an unexpected change of plan. Lieutenant-Commander James Ramage, leading the Dauntlesses of VB-10 Squadron from *Enterprise*, recalled:

The *Enterprise* and Air Group 10 had been waiting a long time for an opportunity to resume action against the Japanese fleet. We particularly wanted to get at their carriers. One rather dismaying event occurred after take-off and rendezvous of my strike group when we received word from the *Enterprise* that the position of the Japanese fleet given to us in our briefing was one degree off in longitude. That meant that the fleet was about 60 miles farther to the west than we had anticipated.

Nonplussed by the sudden addition of 120 miles to their round trip, the raiding forces continued determinedly after their quarry. In front of the Japanese carriers about 40 'Zeke 52s' mounted a spirited air defence in which they shot down six Hellcats and fourteen attack planes for the loss of 25 of their number. The remaining US planes then delivered their attack, in which Avengers torpedoed and sank the small carrier *Hiyo*. Dive-bombers caused damage to the large carrier *Zuikaku* and the small carriers *Chiyoda* and *Ryuho*.

It was dark when the US planes, in many cases flying on the last of their fuel, reached the area where their carriers and escorts were waiting to receive them. To assist the returning crews the warships were ordered to turn on their lights and fire starshell, but the move was not as helpful as it might have been. James Ramage recalled:

> The idea was good, but it was incorrectly executed. Had the order been to turn on lights only on the aircraft carriers, it would have been a great help to us. As it was, it added greatly to the confusion. Every ship was illuminated and it was quite impossible to tell the carriers from other surface ships until one was close aboard. Nevertheless I managed to bring the *Enterprise* group back, augmented by some 30 to 40 stragglers that we had picked up en route, and broke my formation over the *Enterprise* for landing. I made my first pass at the *Enterprise*, but she had a foul deck from a crash landing by another ship's aircraft, so I proceeded to the *Yorktown* and made a very uneventful night landing with only a cupful of fuel remaining.

In the general confusion 80 of the returning planes ran out of fuel and were forced to ditch. Thanks to the warm seas in the area and an effective rescue operation mounted the following day, however, more than three-quarters of the 209 US airmen who came down in the water were picked up alive.

The Battle of the Philippine Sea, or 'The Great Marianas Turkey Shoot' as it came to be nicknamed, was as big a disaster for the Japanese carrier fleet as that suffered at Midway. Of the three large carriers—the only ones that could operate the new Japanese high-performance attack planes—two had been sunk and one had been damaged. One of the smaller Japanese carriers had been sunk and two had suffered damage. And, to crown it all, despite the loss of about three-quarters of their aircraft and crews, the Japanese carrier air groups had inflicted no significant damage on the opposing naval force.

To be successful, carrier-launched air strikes require a particularly high degree of crew training and operational experience. During the action off the Marianas the US Navy air groups possessed those qualities in good measure, while their opponents did not. That, plus the superb performance of the Grumman Hellcat and its presence in large numbers, simply smothered the Japanese attacks. The air action around Task Force 58 off Saipan on the morning of 19 June 1944 was one of the most intensive air actions ever fought. More than 700 carrier planes took part in the fighting, and, of those, 188, or just over a quarter, were destroyed. As has been mentioned, 181 of the losses in the one-sided action came from the Japanese Navy, and these represented 60 per cent of the force it sent into action. The great majority of the losses were inflicted by US fighters, though ships' guns also accounted for a few.

The brilliant defensive action by Task Force 58 inflicted a defeat on the Japanese carrier air groups from which the latter never recovered. The Pacific war would run for another fourteen months, but the Japanese aircraft carriers would play little further part in it.

8

RECONNAISSANCE OVER NORMANDY

From the start of the D-Day invasion of Normandy in June 1944, Allied army commanders received frequent and comprehensive aerial reconnaissance of the enemy-held areas in front of them. In contrast, throughout the initial phase of the land battle their German counterparts received only glimpses of what was happening in parts of the beach-head area. Often the first indication they had of an Allied attack was the preparatory artillery bombardment or when the spearhead units came within view of German forward positions. In an effort to redress this critical deficiency, the Luftwaffe pressed into service a new and completely unproven type of aircraft.

IT WAS NOT that the *Luftwaffe* was short of reconnaissance planes. On D-Day *Luftflotte 3* in France possessed four long-range reconnaissance *Gruppen* with a total of 60 aircraft, for the most part Junkers Ju 88s, Ju 188s and Messerschmitt Me 410s. There was also a tactical reconnaissance *Gruppe* with 42 Messerschmitt Bf 109s with reduced armament and modified to carry cameras. The problem was that the German twin-engine reconnaissance aircraft lacked the performance to operate in the face of the Allied day-fighter patrols and so had to resort to night photographic missions using flares—but night photographs gave far less information than those taken by day.

The Bf 109 reconnaissance fighters flew low-altitude photographic missions over Allied positions by day, but their incursions usually ended with a high-speed dash for home, pursued by defending fighters, and produced nothing like comprehensive cover. These reconnaissance missions were flown at great cost in both aircraft and crews, and produced only a fragmentary picture of the Allied dispositions. The lack of effective aerial reconnaissance brought its inevitable consequences at the end of July 1944, when American troops broke out of the western side of the lodgement area in force and advanced rapidly down the western side of the Cherbourg Peninsula. By the time German field commanders realized what was afoot, it was too late to stop the move or even begin to contain it.

Shortly after the invasion began, *Oberleutnant* Horst Götz was appointed commander of a new air reconnaissance unit that was to

ARADO Ar 234A

Role: Single-seat reconnaissance aircraft.
Powerplant: Two Junkers Jumo 004A jet engines each developing 1,980lb of thrust; two Walter HWK 500A liquid-fuel booster rockets in pods under the outer wings, each developing 1,100lb of thrust for 30sec (once the fuel was exhausted, these pods were jettisoned and fell by parachute, and after landing they were collected for re-use).
Armament: None. The military load carried during the missions described comprised two Rb 50/30 cameras with 50cm focal length lenses, fitted in the rear fuselage.
Performance: Maximum speed 485mph at 19,700ft. Typical speed and height for high-altitude operational photographic run, 460mph at 34,200ft.
Normal operational take-off weight: 17,640lb.
Dimensions: Span 46ft 3½in; length 41ft 5½in; wing area 284 sq ft.
Date of first flight of Ar 234A: August 1943. This version of the aircraft did not enter series production.

operate under the direct control of the *Luftwaffe* High Command (*Versuchstaffel der Oberkommando der Luftwaffe*). Forming at Oranienburg near Berlin, the unit took delivery of the fifth and seventh prototypes of the new single-seat, twin-jet Arado Ar 234A reconnaissance bomber. The aircraft were each fitted with a pair of 50cm focal length reconnaissance cameras in the rear fuselage, arranged to look down and slightly away from each other so as to give overlapping cover on the ground beneath the aircraft. The Arado could fly at 30,000ft at speeds in excess of 450mph, and it was confidently expected that it could penetrate the Allied fighter defences with ease.

Having possession of a couple of advanced aircraft was one thing; getting them, their crews and their supporting personnel to the point from which the unit could fly operational missions over enemy territory was quite another. Spare parts were a constant problem, for the aircraft were hand-made prototypes and these could come only from the makers. There were no proper technical manuals for the Arados and the ground crews had to learn the engineering foibles of the new plane as they went along. Götz and his second-in-command, *Leutnant* Erich Sommer, had to teach themselves to fly the new aircraft (there was no two-seat trainer version) as well as the rudiments of handling its temperamental early-production Jumo 004 jet engines.

Quite apart from the turbojet powerplants, the aircraft had a number of other novel features. To provide the necessary acceleration for it to take-off when carrying a full fuel load, the Ar 234 was fitted with two rocket pods under the wings, each of which developed 1,100lb of thrust for 30 seconds. Once the fuel was exhausted, the pods were jettisoned and fell slowly to earth by parachute. They were then retrieved for re-use.

To meet the stringent range requirement in the *Luftwaffe* specification for the jet aircraft, the Arado design team had been forced to dispense with the conventional-type undercarriage to make space available for extra fuel tanks. The aircraft took off from a wheeled trolley that was released on lift-off and remained on the ground. At the end of the flight the Arado landed on sprung skids that were extended below the fuselage and engine nacelles. The use of the releasable trolley helped to give the Ar 234A a sparkling performance, but at a considerable cost in flexibility of operation. The aircraft could operate only from an airfield that had a trolley available and, in the case of the prototypes, only one trolley had been built for each of them and because of minor differences between the planes their trolleys were not interchangeable. The full significance of this limitation would become clear when the time came to bring the new planes into action.

Ideally Götz should have had at least three months to prepare his unit for action, but the rapid deterioration of the military situation in France ruled that out. On 25 July the two Ar 234As took off from Oranienburg and headed for their operational base at Juvincourt near Reims. On the way Götz's plane suffered an engine failure and he had to fly to the only place where there were both spare engines and a landing trolley—back to Oranienburg.

Erich Sommer reached the French airfield without incident and, after he had landed, the Arado was hoisted on to a low-loader and driven into a hangar to shield it from prying Allied reconnaissance planes. And there, despite the desperate need for its services, the world's most advanced reconnaissance aircraft remained for more than a week (during which the American troops assembled unobserved for their break-out operation). Because of the risk from marauding Allied fighters, the all-important take-off trolley, the rocket pods and other essential items of equipment could not be flown in by transport plane. Instead they were sent to Juvincourt by rail, and despite being accorded the highest priority they took more than a week to get there via the battered French rail network.

Not until the morning of 2 August was the Ar 234 ready for its first combat mission. Perched on its trolley, the strange-looking aircraft was towed to the down-wind end of the airfield. Juvincourt had come under air attack several times, but the bomb craters in the operating surface had been filled in and rolled flat. Sommer climbed into the small cockpit and strapped in. He completed his checks and started the two jet engines, then, when the temperatures had settled down, he released the brakes and eased forward the throttles. Slowly the Arado gained speed as it bumped across the grass surface, then, satisfied that the aircraft was handling properly, the pilot pushed the button to ignite the booster rockets. Sommer felt a reassuring push in the back as the rockets developed full thrust, and its rate of acceleration increased markedly. When the plane reached 100mph he could feel the nose lifting off the ground and the aircraft trying to get airborne, so he pulled the mechanical release for the take-off trolley. Freed of its weight and drag, the aircraft lifted cleanly away. Shortly afterwards the booster rockets exhausted their fuel and the acceleration ceased. The pilot released the two pods and they tumbled clear of the wings. The first-ever jet reconnaissance mission was on its way.

After establishing the aircraft in the climb at 250mph and ascending at 2,500ft/min, Sommer eased the nose round until it was pointing towards the port of Cherbourg. It took the sleek jet twenty minutes to reach its operational altitude of 34,000ft. The view rearwards from the cockpit of the Arado was poor, and from time to time the pilot dropped a wing and turned first to one side then to the other, to glance past his tail to check that there were no enemy fighters in pursuit. Sommer also needed to ensure that the Arado was leaving behind it no tell-tale condensation trail that would have betrayed his position to the enemy. He found that there were no problems on either count.

Once over Cherbourg, Sommer pulled the Arado round on to an easterly heading, eased down the nose until the speed built up to 460mph, then levelled the plane out at 32,300ft. At the flick of a switch the shutters of the plane's two cameras began to click open at 11-second intervals, taking in a swathe of ground just over six miles wide. The day was clear and sunny, and there was scarcely a wisp of cloud to hinder photography as the invasion beaches slid beneath the aircraft. From his vantage point more than six miles high, Sommer could see no sign of the life-and-death struggle taking place on the ground below. If an Allied fighter did attempt to catch the high-flying intruder, it never got close enough for Sommer to see it. The initial photographic run lasted about ten minutes and took in the coastal strip. Then Sommer turned starboard through a semi-circle and flew a second run six miles inland and parallel to the first. Ten minutes after that he began his third run, heading east on a track six miles further inland. Just before he reached the end of the third run the film in the cameras' magazines ran out.

Sommer had finished the most difficult part of the mission but, like any reconnaissance pilot, he knew that the sortie was not completed until he had returned to base with the precious film. As he approached Juvincourt flares were lit to assist him to find the airfield. Keeping a wary eye open for Allied fighters, he made a high-speed descent, extended the skids and plunked the Arado on the grass. Sommer later recalled that on its skids the plane made a very smooth landing on the grass, much smoother than with a conventional wheeled undercarriage on a runway. No sooner had the machine slid to a halt than men began running towards it from several directions. The camera hatch above the rear fuselage was unlocked and opened, then the film magazines were unclipped, lifted out and whisked away for developing. Next, the Arado was hoisted on to its take-off trolley and towed back to the hangar.

In the course of that single flight of about 90 minutes Erich Sommer achieved more than the entire *Luftwaffe* reconnaissance force in the West throughout the previous eight weeks. The 380 photographs he brought back caused an enormous stir, for together they took in almost every part of the Allied lodgement area in Normandy. From Allied records we know that by that date more than 1½ million men and 300,000 vehicles had been put ashore in France. The twelve-man team of *Luftwaffe* photo-interpreters at Juvincourt took more than two days to produce an initial analysis of the photographs, and detailed examination of the prints took weeks.

Soon after Sommer had landed after his epic flight, Horst Götz reached Juvincourt with the other Arado. During the following three weeks the two jet planes flew thirteen further missions, ranging at will over France to photograph their required targets. At last the German field commanders received regular reconnaissance information on what was happening in the beach-head area and on the progress of the American troops advancing deep into France. By 17 August the latter had entered Chartres and Orléans, and were threatening to encircle in the entire German Army Group in Normandy. Under severe pressure, the German troops were forced to withdraw, but then their resistance collapsed and they were plunged into headlong retreat. The time when the Arado's photographs might have a decisive impact on the land battle was past, and now they did little more than provide a detailed picture of a battle that was already lost. On 28 August, with American tanks advancing on Juvincourt, Götz received orders to withdraw his unit to Chievres in Belgium.

During the Battle of France Allied fighters never interfered with the high-flying Arados, and it would seem that the operations of the latter went undetected. Despite a careful search of Allied records this author has found no mention of German jet reconnaissance missions during this period, nor does there appear to have been any reference to them in 'Ultra' decrypted signals. For Erich Sommer and Horst Götz, the continued Allied ignorance of their operations was the greatest compliment that could possibly be paid to the skilful way in which they performed their duties. The task of the reconnaissance pilot was (and, indeed, still is) to penetrate to the target, take the required photographs and return to base with the precious intelligence with as little fuss as possible. If they could do so without the enemy even realizing what had happened, so much the better.

Early in September the first of the new Arado Ar 234B reconnaissance aircraft was delivered to Götz's unit. This version was fitted with a normal tricycle undercarriage, which made for much greater flexibility of operation and allowed the plane to operate from almost any airfield. The new version was 24mph slower than prototypes and its radius of action was somewhat less, but despite these drawbacks the new Arados ranged far and wide over Allied territory with little interference for the remainder of the war.

Had the Arado Ar 234As been available to fly reconnaissance missions a couple of months earlier, in time for D-Day, there is little doubt that with their photographs German field commanders would have deployed their forces much more effectively than was the case. And, goodness knows, even without their help the German Army put up an extremely hard fight in Normandy. As an old military adage assures us, 'Time spent on reconnaissance is seldom, if ever, wasted.'

BOMBERS AGAINST THE *TIRPITZ*

The final generation of battleships made extremely difficult targets for air attack. Although they were large and difficult to hide, they were also very well armed and had strong armour protection. If these vessels were in port or at an anchorage they usually enjoyed the additional protection of smoke-generating units, anti-torpedo nets, anti-aircraft guns and, usually, shore-based fighters. To stand a reasonable chance of destroying such a target it was necessary to use either a very large number of planes carrying conventional armour-piercing bombs, or a small number of planes equipped with more specialized types of weapon. During its final series of attacks leading to the destruction of the battleship Tirpitz, *the Royal Air Force adopted the latter course . . .*

THE GERMAN battleship *Tirpitz* spent almost her entire life holed up in the Norwegian fjords. Too powerful to be ignored, she was a constant menace to the Allied convoys carrying supplies and equipment around the north of Norway to assist the Soviet Union's war effort. *Tirpitz* rarely put to sea, and when she did she saw relatively little action. But by her very presence she tied down a large number of Allied warships, including battleships, that could have been better used elsewhere.

Tirpitz, sister-ship of the famous *Bismarck*, was one of the most powerful warships of her time. With a displacement of 42,900 tons, she was 7,900 tons heavier than the *King George V* class battleships, the nearest equivalent in the Royal Navy. Much of the difference in weight was accounted for by the German ship's extra armour. To give protection against naval gunfire and torpedoes, *Tirpitz* had a vertical belt of side armour 12.6in thick that extended from 8ft below the waterline to a similar distance above it; above that belt extending to deck level was a vertical belt of armour 5.7in thick. To keep out aerial bombs, the vessel carried a layer of horizontal armour just over 3in thick to protect the machinery spaces, increasing to nearly 4in above the ammunition magazines. Where the deck armour met the side armour, the former was inclined at an angle to keep out plunging shot. The German battleship had an abnormally wide beam of 118ft, giving her a greater degree of initial stability than any comparable ship in the British or US Navies (battleships built for the latter had to be narrow

BATTLESHIP *TIRPITZ*

Surface displacement: 42,900 tons.
Armament: Eight 15in (38cm) guns main armament; twelve 5.9in (15cm) guns secondary armament; sixteen 4.1in (10.5cm) high-angle anti-aircraft guns; sixteen 37mm and fifty-eight 20mm short-range automatic anti-aircraft guns; eight 21in torpedo tubes.
Performance: Maximum speed 30kts; range 9,000 miles at 19kts.
Dimensions: Length overall 822ft; beam 118ft.
Complement: 2,400.
Date operational: January 1942.

enough to pass through the locks of the Panama Canal, but for obvious reasons that was not a design requirement for German warships).

British air and naval forces made numerous attacks on *Tirpitz*. The first, when she was located at sea off Norway on 9 March 1942, was by a dozen Albacore torpedo bombers launched from HMS *Victorious*. The German battleship headed into the wind at maximum speed, and the slow biplanes lacked the necessary speed advantage to carry out a co-ordinated attack on her. Several torpedoes were aimed at the rapidly manoeuvring warship but none of them hit. The night attacks by Royal Air Force bombers on 31 March and 28 and 29 April of that year, when the ship was in Fötten Fjord near Trondheim, were no more successful.

On 22 September 1943 midget submarines of the Royal Navy crept into her anchorage, then in Kaa Fjord at the north of Norway, and planted four 2-ton mines on the sea bed beneath the vessel. The detonation of two of the weapons caused widespread internal damage which required repairs lasting several months. Such an attack depended on its novelty to have any chance of success, and could be mounted only once.

On 3 April 1944 a large force of Fleet Air Arm Barracudas dive-bombed the battleship and obtained twelve hits. Most of the bombs were released from too low an altitude, however, and they failed to gain sufficient velocity to penetrate the ship's armoured deck. Although some damage was caused to the superstructure, the battleship was ready for sea again within a month. The smokescreens around *Tirpitz* nullified similar attacks by the Fleet Air Arm on 17 July and 22 August, limited that on 24 August to two hits which caused relatively little damage, and ruined the attempt on 29 August.

In September 1944 it was again the turn of the Royal Air Force to pit its strength against *Tirpitz*. A force of thirty-eight Lancasters from Nos 9 and 617 Squadrons deployed to Archangel in northern Russia for the attack, equipped with two novel types of weapon: the 12,000lb

'Tallboy' bomb and the 500lb 'Johnny Walker' underwater 'walking' mine.

Designed by famous inventor Barnes Wallis, the 'Tallboy' had first been used in action less than four months earlier. It was 20ft long and 3ft in diameter, and it took up the whole of the Lancaster's bomb bay. The weapon contained 5,100lb of Torpex explosive and was designed to penetrate deep into the ground before exploding, aiming to cause an 'earthquake' that would literally shake the target to pieces. The four fins at the rear of the weapon were angled at 5 degrees to the airflow, to cause the bomb to rotate about its longitudinal axis during its fall. This rotation provided spin stabilization to hold the bomb straight as its velocity approached the speed of sound. The 'Tallboy' had not been designed for use against warship targets, but nobody doubted that a direct hit or a near-miss from one of these weapons would inflict serious damage on *Tirpitz*.

The 'Johnny Walker' or 'JW' mine was purpose-designed for use against ships at anchor. The weapon weighed approximately 400lb and, before use, lead was added to adjust its weight so that it would have neutral buoyancy in the area of sea in which it was to be dropped. The small size of the 'JW' (6ft long, 15¼in in diameter) meant that the Lancaster's bomb bay could accommodate a dozen of these weapons. In both its appearance and its concept, the 'JW' could scarcely have been more different from 'Tallboy'. Whereas the latter depended on sheer brute force to smash its way through the armoured deck and into the vitals of the ship from above, the 'JW' employed more subtle methods. The latter was designed to attack *from the vulnerable underside* where, in the case of *Tirpitz*, the plating was only two-thirds of an inch thick.

The method of operation of the 'JW' was as follows. After its release from the aircraft, the mine descended on a 4ft-diameter drogue parachute. If it struck a solid surface (i.e. the ground or the superstructure

---

AVRO LANCASTER I

Role: Seven-seat heavy bomber.
Powerplant: Four Rolls-Royce Merlin 24 12-cylinder, liquid-cooled, inline engines each developing 1,640hp at take-off.
Armament: (Offensive, during the attack on *Tirpitz* on 15 September 1944) One 12,000lb 'Tallboy' bomb or twelve 'Johnny Walker' mines; (defensive) eight Browning .303in machine guns, two each in the nose and mid-upper turrets and four in the tail turret.
Performance: Maximum speed 287mph at 11,500ft; normal cruising speed 220mph at 20,000ft; service ceiling 24,500ft.
Normal operational take-off weight: 68,000lb.
Dimensions: Span 102ft; length 69ft 6in; wing area 1,297 sq ft.
Date of first production Lancaster I: October 1941.

---

of a ship), an impact fuse detonated the warhead. If the mine entered water, the drogue was released and the weapon sank. When the mine reached a depth of 60ft during its descent, a hydrostatic valve released a flow of hydrogen from a high-pressure container in the rear of the weapon and passed it, via a pressure-reducing valve, to eject the water out of the buoyancy chamber in the nose. The weapon became lighter than the surrounding water, its descent ceased and it started to rise nose-first. The fins on the sides of the weapon were so arranged that when the 'JW' ascended or descended through water it travelled at an angle of about 30 degrees to the vertical, thus giving a horizontal displacement of about 30ft during each such move. When the nose of the weapon was uppermost, the nose fuse was made live so that if the weapon struck anything hard anything during its ascent (i.e. the bottom of a ship) the warhead would detonate.

The 'JW' warhead contained 100lb of Torpex/aluminium formed into a shaped charge, with a concave area at the front lined with soft metal. The explosive was detonated from the rear, so the force of the explosion was focused on the metal liner. The latter would dissolve into a slug of molten metal that was hurled forwards from the warhead at very high velocity. This slug had enormous penetrative power and, once it inside the bowels of the ship, it could cause severe local damage. That effect would be compounded by the ingress of water through the hole made by slug. Powerful though these effects were, they were only part of the damage mechanism the weapon was designed to produce. The inclusion of powdered aluminium in the explosive charge meant that when it detonated large amounts of gas were generated. The huge bubble thus formed was capable of lifting the ship suddenly at the point of detonation. Then, an instant later, the bubble collapsed, causing that part of the ship to drop several feet. The sudden up and down movement imposed severe forces on the hull of a vessel and could break its back (sea mines produce a similar effect).

If the 'JW' reached a depth of 15–20ft in its ascent it had obviously missed its target (the bottom of *Tirpitz* was about 34ft below the surface). At that depth the hydrostatic valve stopped the flow of hydrogen and a port opened to release the gas in the buoyancy chamber. Water flooded into the chamber and the mine, now heavier than the surrounding water, turned over and began to descend nose-first at an angle of 30 degrees. During the descent the nose fuse automatically became 'safe', so that if the weapon struck the sea bed it would not detonate. When the mine approached a depth of 60ft the hydrostatic valve turned on the supply of hydrogen again, the water was forced out of the buoyancy chamber and the cycle was repeated. The container in the rear of the weapon had sufficient hydrogen for the 'walking cycles' to continue for up to an hour. If at the end of that time the 'JW' had not found a target, the self-destruct mechanism set off the warhead to prevent the secret weapon falling into enemy hands.

The attack on *Tirpitz* was launched on 15 September, when 27 Lancasters took off from Archangel. Twenty of the aircraft were each loaded with a single 'Tallboy', six planes carried twelve 'JW' mines each, and the remaining Lancaster was to film the attack for later analysis.

Wing Commander J. Tait, the CO of No 617 Squadron, had drawn up the attack plan. The 'Tallboy' aircraft were to bomb first, running in from the south. As is usually the case, the attack plan had to take into account a number of sometimes conflicting factors. The time of flight of the 'Tallboy', from the planned release altitude of 11,000ft to impact, was about 26 seconds, and when it detonated it would hurl debris and water high into the sky. The explosions from several such bombs, and smoke from any resultant fires, might conceal the target from any aircraft still running in to bomb. This, and the need to prevent German anti-aircraft gunners concentrating their fire on individual planes to disrupt their bombing runs, required a concentrated attack if possible with the last aircraft releasing its bomb within 26 seconds of the leader. On the other hand, the bomb-aimers were all trying to hit the same aiming point, the middle of the battleship, so during the final part of their bomb run there would be several planes flying on the same or on converging headings. If the aircraft bunched together too tightly there would be a risk of collisions, or of the bomb run being disrupted, if aircraft flew into the turbulent slipstream from those in front.

Tait's plan was for the aircraft to attack in four V-shaped waves each of five aircraft. Waves of bombers were to follow each other at intervals of about 800yds; thus the entire formation was 1½ miles long and at its attack speed of 230mph would take about 22 seconds to pass over the target. This ensured that the last Lancaster would have released its bomb several seconds before that from the leader detonated, giving all the aircraft an unobscured target at which to aim. It also meant that the German gunners could not afford to concentrate their fire on individual planes and allow others in the force to bomb unhindered. To keep the aircraft out of each other's way during the bombing runs, there was a 50ft altitude separation between adjacent bombers in each wave. Furthermore, succeeding waves were stepped up by 1,000ft to keep each out of the slipstream of the one ahead of it. During this attack the lowest aircraft in the first wave released its bomb from 11,350ft and the highest in the final wave released from 17,500ft.

The six 'JW' aircraft attacked a few minutes later, running across the fjord in two waves from south-east to north-west and releasing their loads from altitudes between 10,000 and 12,000ft. This order of attack was necessary because it was essential that the mines enter the water after the last 'Tallboy' detonated, or the blast from the latter was likely to cause 'countermining'.

By the time the Lancasters reached the Kaa Fjord the German smokescreen was well developed, however, and, as a result, the Lancasters' bomb-aimers had to aim their 'Tallboys' at where they thought the battleship lay, or at the muzzle flashes visible through the smoke. Some of the bombers failed to release on their first run and made a second attack run; fortunately for these crews, the smokescreens that hid the battleship from the Lancasters also hid the Lancasters from the anti-aircraft gunners. No German fighters attempted to interfere with the attack, and none of the raiders suffered any damage.

There can be little doubt that the smokescreens saved *Tirpitz* from annihilation that day. Only one 'Tallboy' struck the battleship a glancing blow, yet its effect was devastating. The bomb hit the starboard side of the foredeck between the bow and the forward gun turret, passed through the flare of the ship's side and detonated in the water a few feet below and to one side of the ship. The explosion blew out a large area of bow plating below the waterline, to a distance of about 55ft. There was extensive flooding throughout almost the entire bow area and, with the admission of about 1,500 tons of water during a counter-flooding operation to restore the ship to an even keel, the draught of *Tirpitz* was increased by more than 8ft. The shock of this and several other near-misses affected equipment throughout the vessel; for example, the severe vibration caused damage to the main engines, and most of the optical range-finding instruments were put out of action. Although still afloat, the battleship was barely seaworthy, and she was certainly in no condition to fight.

The 'brute force' method of attack with the 'Tallboy' had obviously been a success. But what of that by the more sophisticated 'Johnny Walker' mines? Had one of these detonated against the battleship's hull, the weapon would have been accorded the fame given to the bouncing bomb that broke the German dams. But it was not to be. During the next hour the 'JW' mines wandered aimlessly and silently up and down the fjord. Then, as the last of the hydrogen gave out, one by one they blew themselves up harmlessly. The completeness of their failure had but one redeeming factor: it preserved the secrecy of the operation of the weapon, leaving the German Navy with no inkling of what had been attempted.

After the attack German Navy engineers estimated that to restore *Tirpitz* to her full operational capability would require the rebuilding of much of the bow section. That would take a minimum of nine months' uninterrupted repair work in a fully equipped dock, which meant taking her back to Germany. In the current war situation this was not considered feasible, and it was decided to carry out makeshift repairs and move the ship to Tromsø Fjord some 200 miles to the south. There the battleship was to serve as a floating gun battery to stiffen the land defences of the area. The German Navy went to great pains to conceal the true extent of the damage to *Tirpitz*, recognizing that so long as she was perceived as a threat, the battleship would continue to tie down substantial Allied naval forces. It took several weeks for information on the damage to filter through to Royal Navy Intelligence, and in the meantime the battleship remained a high priority target.

The move to Tromsø Fjord took place in mid-October. It was to seal the fate of the battleship, for the new location lay just within the radius of action of Lancasters operating from bases in Scotland. A number of these aircraft were fitted with overload tanks to bring their fuel load to 2,406 Imp. gallons, and with more powerful engines. To compensate for the extra weight, the mid-upper turrets and the guns and ammunition from the front turrets were removed from the aircraft, as were all items of equipment not considered essential for the mission.

By the final week in October these modifications were complete, and on the 29th a force of 37 Lancasters took off from Lossiemouth for another attack on the battleship. Now the vessel was moored in water too shallow for the 'JW' mine to operate, so all the aircraft carried 'Tallboys'. The German Navy had not had time to move the smoke-generating equipment to Tromsø, but on this occasion it did not matter. The aircraft arrived over the target to find the warship enshrouded in cloud. Thirty-two crews aimed their bombs into the area where the *Tirpitz* was thought to be, but no hits were claimed. In fact one weapon detonated about 80ft from the ship's port side, causing damage to a propeller shaft and a rudder, but that was all.

Soon after the attack Royal Air Force Intelligence learned that the *Luftwaffe* had moved a *Staffel* of FW 190 fighters to the airfield at Bardufoss, 40 miles from Tromsø, from where they might interfere with an attack on *Tirpitz*. The battleship was still a high-priority target and, if the next attack were to succeed, it was essential that the RAF retain the element of surprise for as long as possible before the force reached the target. As part of the general effort to follow the continual changes in the German radar chain, elint (electronic intelligence) planes had plotted the locations and arcs of cover of the stations along the coast of Norway. These were positioned to give warning of aircraft approaching any part of the coast at altitudes of 5,000ft or above. But if a raiding force remained at 1,500ft or below, there was a chink in the curtain mid-way up the coast through which the planes might pass unseen. It was planned that during the next attack the bombers would fly through the hole in the radar cover at 1,500ft and continue heading east and climb over the mountains into neutral Sweden. The planes were then to head north- east, keeping the mountain barrier between themselves and the German radar stations, until they began the climb to enter their bombing runs.

On 12 November a force of 29 Lancasters took off to attack the battleship, following the route as planned. Some of the bombers were detected crossing the coast, however, and the report was flashed to *Tirpitz* that yet another attack might be in the offing. What the RAF planners had not known was that the fighters at Bardufoss posed no real threat to the attackers. The unit had traded its worn-out Messerschmitt Bf 109s for Focke Wulf 190s only a few days earlier and its pilots, many of whom were straight from training and had little flying experience, needed a period of conversion before they were proficient on their new mounts. The remit of the *Staffel* commander, *Oberleutnant* Werner Gayko, was that, except in an emergency, his unit was responsible only for the defence of the airfield and the immediate area surrounding it. So far as Gayko was concerned no emergency existed, for he had received no request to go to the assistance of the battleship.

As they passed over the distinctive shape of Lake Torneträsk, the stretch of water 1,125ft high in northern Sweden, the bombers moved into attack formation and their pilots pushed on full throttle to commence the climb to attack altitude.

At 9.05 a.m. the formation of bombers hove into view of the look-outs on *Tirpitz*. By then the battleship was fully closed up for action, her watertight doors shut and her weapons loaded and pointing skywards. Even the four 15in guns in the two forward turrets were ready for action, their barrels raised to maximum elevation and aligned on the raiders. As the Lancasters came within range of the big guns, 13 miles, the weapons hurled their one-ton shells in the direction of the planes. The deafening crashes of the guns echoed and re-echoed from the mountain walls surrounding the fjord, then the barrels began lowering smoothly to the horizontal for re-loading. In succession four large clouds of black smoke appeared in the sky near to the bombers, but not so near as to harm any of them. As the range closed, the battleship's 10.5cm guns joined in the action, followed soon afterwards by the 3.7cm automatic weapons' strings of red or green tracers.

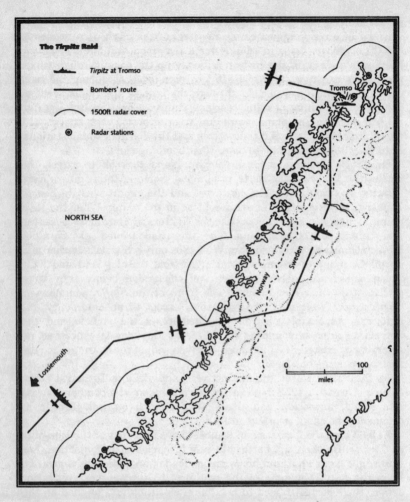

The Tirpitz Raid

- ⚓ *Tirpitz* at Tromso
- —— Bombers' route
- ⌒ 1500ft radar cover
- ◎ Radar stations

NORTH SEA

Tromso

Norway

Sweden

Lossiemouth

0   100
      miles

Tait's tactics for this attack were the same as those used during the previous two, with planes attacking in closely spaced waves stepped up in altitude. And this time, as luck would have it, there was neither cloud nor a smokescreen to hinder the attackers. The first wave of Lancasters reached the bomb-release point and in unison the five aircraft disgorged their 'Tallboys'. Falling slowly at first but rapidly gathering speed, the salvo of huge bombs bore down on the battleship. Seven seconds later the second wave bombed, then the third, then the fourth wave.

Three 'Tallboys' scored direct hits on the ship. Surprisingly, one of the bombs failed to cause serious damage; it ricochetted off the armoured deck near to 'B' turret and, breaking up and spilling burning charge, it skidded away. The nose of the weapon was later found on a mud bank 200yds from the impact point. The catastrophic damage was caused by the other two weapons, which smashed into the ship within 20yds of each other on the port side, amidships. These penetrated the hull of the ship, whereupon they detonated. In combination with one or more near-misses at about the same time, the explosions tore a huge hole about 200ft long in the port side.

Hundreds of tons of water flooded into the battleship and, despite counter-flooding to correct the list to port, the latter increased steadily until it reached 20 degrees. There the ship steadied for a while, with one bilge keel jammed hard against a bank of mud on the sea bed, before the list slowly resumed. Below decks severe fires were raging. One of them reached a magazine which exploded, lifting the hefty 'C' turret off its mounting and hurling it over the side. The list continued relentlessly until the ship ended up with her superstructure embedded in the mud on the bottom of the fjord and with part of her hull showing above the surface. From the beginning of the attack until the ship came to rest, it had taken about ten minutes. Of the 1,500 sailors on board the warship at the time, about half had been killed.

In the Royal Air Force there was considerable jubilation at the destruction of the German Navy's last remaining modern battleship by air attack. The action proved that no warship, no matter how large, how well armoured or how well protected by anti-torpedo nets, could survive the punishment that could be delivered from the air. The triumph was in no way muted by a carping comment from a senior Royal Navy officer, who tried to denigrate the junior Service's success by pointing out the literal but quite irrelevant fact that *Tirpitz* had not been sunk because part of her hull (the bottom) was still protruding from the water. The aviators could afford to laugh off the crass remark. Given the greater truth that the big bomber was now the master of the battleship, it mattered not one jot if any part of the latter remained above water at the end of the attack.

CONFOUND AND DESTROY

*At the end of 1943 Royal Air Force Bomber Command gained a new capability
to assist its night raiders: No 100 Group, a force with the task of reducing the
effectiveness of the German air defences. The Group's commander, Air
Vice-Marshal Edward Addison, planned a two-pronged attack on the defences.
First, by jamming and spoofing the German radar systems and blocking the
radio channels, he would make it more difficult for night fighters and
anti-aircraft gunners to find and engage the bombers. Second, by sending
long-range night fighters to seek out their German counterparts and attack
their airfields, the Group would impose constraints on the operations of the
defending night fighter force. In recognition of the dual-nature of its role, the
Group's official motto was 'Confound and Destroy'.*

D URING THE AUTUMN of 1944 No 100 Group began to make
its presence felt during the night attacks on targets in
Germany. The role of the formation was termed 'bomber
support' (what is now called 'defence suppression'). No 100 Group's
jamming force comprised four squadrons of converted heavy bombers:
No 171 with Halifaxes, No 199 with Stirlings (later it would convert to
Halifaxes), No 214 with Flying Fortresses and No 233 with Liberators
(later it would convert to Flying Fortresses). These aircraft were
modified to carry a menagerie of specialized electronic jamming equipment: 'Mandrel' and 'Carpet' to jam the Germans' ground radars,
'Piperack' to jam their night fighters' airborne interception radars and
'Jostle' to jam their communications radio channels. In addition,
several of the aircraft were modified to carry large quantities of
'Window' metal foil to create thousands of false targets on the enemy
radar screens to distract and confuse the defenders.

The Group's destroying element comprised seven squadrons of
Mosquito night fighters. In addition to their normal airborne interception radar, some of these aircraft carried the 'Serrate' or 'Perfectos'
homing devices. 'Serrate' picked up the transmissions from the
German night-fighter radars and gave a bearing that enabled the
Mosquito crews to home on their source. 'Perfectos' was even cleverer:
it radiated signals to trigger the IFF identification equipment of any
German aircraft within a range of about fifteen miles, and the latter's
coded reply signal betrayed its exact whereabouts. 'Perfectos' provided

the three pieces of information necessary for a successful interception: it gave relative bearing and distance, as well as providing a positive hostile identification of the aircraft under observation. The last of these was particularly valuable, since the Mosquitos were to deep in enemy territory where there might be a few German night fighters in an area of sky filled with several hundred friendly bombers. No longer could *Luftwaffe* night fighters cruise over their homeland concerned only with finding and shooting down bombers: now these hunters were liable at any time to become the hunted.

As well as seeking out German night fighters in the air, two of the Mosquito units, Nos 23 and 515 Squadrons, specialized in flying night intruder missions against the enemy night-fighter bases. These aircraft would orbit over the enemy airfields for hours on end, and bomb or strafe any movement seen on the ground.

To provide elint support for these operations the Group had its own 'Ferret' squadron, No 192, with Halifaxes, Wellingtons and Mosquitos fitted with special equipment to collect signals from the various German radar systems.

No 100 Group's jamming element flew four general types of mission in support of the night bombers: the 'Mandrel' screen, the 'Window Spoof', the Jamming Escort and the Target Support operation. The 'Mandrel' screen usually involved between ten and sixteen aircraft orbiting in pairs along a line just clear of enemy territory, with an interval of fifteen miles between pairs. Each aircraft carried several 'Mandrel' transmitters, and the purpose of the operation was to produce a wall of jamming about 100 miles long, to prevent the German early-warning radar operators from seeing aircraft movements behind the screen. Usually the 'Mandrel' screen was employed to conceal the approach of a raiding force, but when no raid was planned it was erected to cause the enemy controllers to think that

---

CONSOLIDATED LIBERATOR B.VI (B-24H)

Role: Ten-seat jamming support aircraft.
Powerplant: Four Pratt & Whitney R-1830 14-cylinder, air-cooled, radial engines each developing 1,200hp at take-off.
Armament: Defensive armament comprised four Browning .5in machine guns, two each in powered turrets in the tail and above the fuselage. Sometimes the aircraft carried a small number of target-indicator bombs to give realism to their feint attacks. They carried several items of radar and radio-jamming equipment and a large quantity of 'Window' was required for some types of operation.
Performance: Maximum continuous cruising speed 278mph at 25,000ft
Normal operational take-off weight: 56,000lb.
Dimensions: Span 110ft; length 67ft 2in; wing area 1,048 sq ft.
Date of first production B-24H: June 1943.

---

a raid was in the offing and so force them to scramble night fighters to waste their dwindling supplies of aviation fuel.

The 'Window Spoof' comprised up to twenty-four aircraft in two formations of twelve aircraft flying in line abreast, with 2¼ miles between the aircraft and the second line some 30 miles behind the first. Each aircraft released 'Window' at a rate of thirty bundles a minute, one every two seconds. In this way the formation could produce on enemy radar the illusion of a bomber stream of some 500 aircraft. The aim of the tactic was to lure night fighters away from the real raiding forces (each real bomber stream also dropped 'Window', though at a lower rate, so that it was impossible to tell the real attacks from the feints).

The Jamming Escort role involved Fortresses and Liberators of No 100 Group flying above the main bomber stream and jamming with 'Jostle' and 'Piperack' transmitters to blot out, respectively, the German night fighters' radio channels and their airborne interception radars. When the Jamming Escort aircraft arrived at the target they often assumed the Target Support role, in which they orbited in the area and operated their 'Carpet' transmitters to jam the frequencies used by the German flak control radars.

Sergeant Kenneth Stone, an air gunner of No 233 Squadron flying Liberators, described his impressions of the types of operations his unit flew:

> 'Window Spoof' raids were carried out by a few aircraft dropping the metal foil to simulate a large force of aircraft on the enemy radar. The operation was very precisely worked out; there would be a rendezvous point in a safe area and timing was critical to within two minutes. If a crew arrived later than this it had to abort the mission because one aircraft late was a sure give-away on radar. All aircraft had to go in together and dispense 'Window' at a regular pace. The aircraft flew a 'corkscrew' course to disperse the 'Window' more effectively and give the illusion of a larger force than was actually present. Generally six spoofers would show up on radar as 300-plus plots. A spoof target was generally selected which either committed the defences in that area and left the genuine target free, or at worst split the defences and rendered a proportion of them useless. There were two methods of ending this type of spoof. Either we continued to the spoof target and dropped target markers there; or we stopped 'Windowing' short of the target and dived away at a great rate of knots. There is no doubt that the 'Window Spoofs' were highly successful in achieving their purpose, by fooling the defences and diverting their effort. The casualty rate for spoofing aircraft was not so high as might have been expected, because the enemy night fighters would be tied up with the paper-chase and could not winkle out the real aircraft.

In Stone's view the Target Support operation was the most dangerous of those flown by his unit, and in recognition of the hazards these missions were shared out amongst the crews in strict rotation:

> The general principle was to cover the target from five minutes before the initial marking began until five minutes after the bombing stopped. The big hazard was having to hang around while the bomber boys ran in, bombed, and got the hell out of it! Fifteen minutes seemed a long, long time suspended over the inferno below. The support aircraft generally flew some 2,000 to 4,000 feet above the bomber stream and jammed the flak and searchlight radars, the night-fighter R/T frequencies, the night fighter radars, etc.—in other words diverting the defensive forces away from the bombers during the most critical period.

The reader may gain an impression of the way in which No 100 Group's tactics dovetailed with those of the rest of Bomber Command from a more or less typical operation of the late war period, that on the night of 20/21 March 1945. The targets were the oil refineries at Bohlen near Leipzig and at Hemmingstedt near Hamburg; the former was to be attacked by 235 heavy bombers, the latter by 166.

The first action by Bomber Command that night was a large-scale nuisance raid on Berlin by 35 Mosquitos, beginning at 9.14 p.m. The Mosquitos, flying fast and high, required no support from No 100 Group's jamming force. As the raiding force moved in, night-fighter Mosquitos of Nos 23 and 515 Squadrons fanned out over Germany making for the enemy night-fighter bases likely to become active that night. When the intruders reached their objectives they orbited, waiting to pounce on any aircraft seen taking off or landing.

Just after 1 a.m. the main raiding force bound for Bohlen crossed the French coast and headed south-east towards southern Germany. Also heading across France, on a track almost parallel to that of the Bohlen attack force but a little further south, was a feint attack force of 64 Lancasters and Halifaxes. These aircraft belonged to operational

---

DE HAVILLAND MOSQUITO NF.30

Role: Two-seat night fighter.
Powerplant: Two Rolls-Royce Merlin 76 V12, liquid-cooled, inline engines each developing 1,535hp at take-off.
Armament: Four Hispano 20mm cannon mounted in the lower part of the nose.
Performance: Maximum speed 407mph at 28,000ft; climb to 15,000ft, 7min 30sec.
Normal operational take-off weight: 20,000lb
Dimensions: Span 54ft 23in; length 49ft 10¾in; wing area 454 sq ft.
Date of first production Mosquito NF.30: March 1944.

JUNKERS Ju 88G-6

Role: Three-seat night fighter.
Powerplant: Two Junkers Jumo 211J inline engines each developing 1,410hp at take-off.
Armament: Four MG 151 20mm cannon mounted under the fuselage; some aircraft carried two further MG 151 cannon in an oblique installation firing upwards and forwards.
Performance: Maximum speed 389mph at 30,000ft; initial rate of climb 1,655ft/min.
Dimensions: Span 65ft 10½in; length 52ft 1½in; wing area 587 sq ft.
Date of first production Ju 88G-6: Summer 1944.

conversion units and were flown by crews in the final stages of their training.

No 100 Group's electronic trickery began at 2.05 a.m. on the morning of the 21st. Established in a line 80 miles long over France and just inside Allied-held territory, seven pairs of Halifaxes of Nos 171 and 199 Squadrons turned on their 'Mandrel' equipment to provide a wall of jamming to conceal the approach of the Bohlen attack force, which by then had split into two separate parts.

Running across France outside the cover of the 'Mandrel' screen, the feint attack flown by trainee crews continued heading east in full view of the enemy radars. German night fighters moved into position to block the threatened incursion but, at 2.55 a.m., when the bombers were just short of the German border, the feint attackers turned round and went home. A few minutes later, well to the north, the two Bohlen attack forces burst through the 'Mandrel' jamming screen and crossed the Rhine into German-held territory. Twenty miles ahead of the bombers flew four Halifaxes of No 171 Squadron and seven Liberators of No 223 Squadron dropping 'Window' to conceal the strength of the attacking forces. Flying ahead and on the flanks of each of the bomber streams, 33 Mosquito night fighters of No 100 Group began their deadly game of hide-and-seek with their enemy counterparts.

Shortly before 3 a.m. a Mosquito of No 85 Squadron, with pilot Flight Lieutenant Chapman and radar operator Flight Sergeant Stockley, picked up IFF identification signals on 'Perfectos' from an enemy aircraft at a range of 12 miles. Chapman afterwards reported:

At 0255 hours just after passing Hamm on the way in to escort the bomber stream we got a 'Perfectos' contact at 12 miles' range—height 12,000 feet. Range was closed to 1 mile but no AI [radar] contact was obtained and the range started to increase again, so deciding that the contact must be below we did a hard diving turn to port down to 9,000 feet and finally D/F'd [took a bearing] on to the target's course at 7 miles'

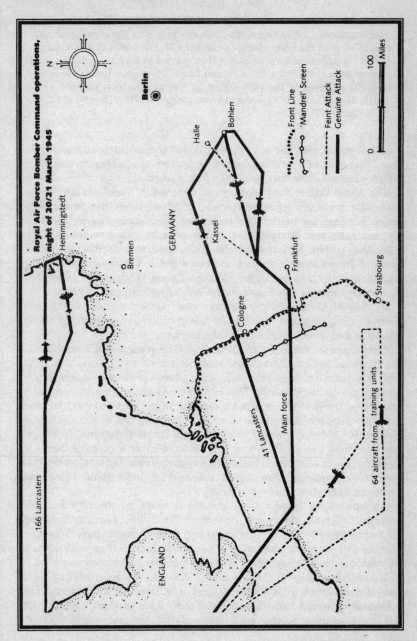

Royal Air Force Bomber Command operations, night of 20/21 March 1945

N

Berlin

Halle

Bohlen

Kassel

Frankfurt

Strasbourg

GERMANY

Bremen

Hemmingstedt

Cologne

ENGLAND

166 Lancasters

41 Lancasters

Main force

64 aircraft from training units

Front Line

'Mandrel' Screen

Feint Attack

Genuine Attack

0   100
         Miles

range. We closed to 6 miles' range on a course of 120° and an AI contact was obtained . . . The target was still climbing straight ahead and was identified with the night glasses as an Me 110. I closed in to 600 feet and pulled up to dead astern when the Hun started to turn to port. I gave it ½ ring deflection with a three-second burst whereupon the E/A [enemy aircraft] exploded in the port engine in a most satisfying manner with debris flying back. It exploded on the ground at 0305 hours, position 25–30 miles NW of Kassel.

From start to finish the engagement lasted ten minutes and, as can be seen, this type of operation tended to involve the Mosquito crew in a lengthy chase before they reached a firing position.

The spoof tactics that night were successful. The German fighter controller seriously underestimated the strength of the two raiding forces heading for Bohlen; he estimated their strengths as about 30 aircraft each and thought they might even be 'Window' feints. Only after the raiders had crossed the Rhine and reports had started to come in from German ground observers did it become clear that the southerly force was far larger than had been thought: no amount of electronic jamming could conceal the roar of 800 aircraft engines.

By now 89 German night fighters were airborne and orbiting over their holding beacons, waiting for their controller to clarify the air situation and direct them against the bombers.

That night the intention of the No 100 Group operation was to create the impression that the raiders' main objective was Kassel, and soon after the main raiding force had crossed into German-held territory it turned north-east towards the city. The German fighter controller swallowed the bait and ordered almost all of his night-fighters to head for radio beacons *Silberfuchs*, *Werner* and *Kormoran* in the vicinity of Kassel. He ordered the rest of his force, a single *Gruppe* of Ju 88s, to move to radio beacon *Otto* near Frankfurt to cover a possible threat to that city. Soon afterwards there came reports from Kassel that the city was under imminent threat of attack, as Pathfinder flares blossomed overhead and a few bombs detonated.

German night fighters were ordered to move on the city, but this was no full-scale onslaught, merely a feint by Mosquito bombers backed by No 100 Group Liberators and Halifaxes dropping 'Window'. During the course of this spoof a German night fighter shot down a Liberator of No 233 Squadron; only one member of the crew survived.

Meanwhile, some 25 miles south of Kassel, the main raiding force had turned away from that city and was heading for Bohlen. The Liberator crewmen had not sacrificed their lives in vain, however, for the feint against Kassel kept most of the German night fighters uselessly in that area for nearly half an hour. Not until 3 a.m. did the German fighter controller realize that he had been tricked and order his force to head east in pursuit of the raiders. Six minutes after that he gave the probable target as Leipzig, the city nearest Bohlen, but by then the vanguard of the raiding force was within thirty miles—eight minutes' flying time—of the target.

Still No 100 Group had not exhausted its repertoire of tricks. Just short of Bohlen six Flying Fortresses and Halifaxes broke away from the main raiding force and ran a 'Window' trail to the important oil refinery complex at Leuna which lay twenty miles to the north-west. Twelve Lancaster bombers accompanied the jamming aircraft to give substance to the spoof. When the feint attackers arrived over the complex they dropped further target markers and the Lancasters put down their loads of bombs. Leuna lay directly in the path of the German night fighters streaming to the east, and the spoof attack delayed their arrival at the real target still further. One Lancaster crew paid the supreme price for the precious minutes of additional delay inflicted on the defending night fighters in reaching the main target.

The 211 Lancasters assigned to the Bohlen raid reached their objective and carried out a concentrated eleven-minute attack. The five Flying Fortresses and the Liberator that had provided Jamming Escort Support along the route to the target now orbited over the refinery in the Target Support role throughout the period of the attack. Not until 4.10 a.m., as the last raiders were leaving Bohlen, did the first of the German night fighters arrive in the area. Their radar operators encountered severe jamming and they had great difficulty picking out their prey amongst the large numbers of 'Window' returns. To add to the defenders' confusion, as the Bohlen attack force withdrew to the west, the No 100 Group Halifaxes that had operated in the 'Mandrel' screen role had a further part to play. They now ran a further 'Window Spoof' 'attack' on Frankfurt, and dropped target markers to simulate the opening of a large-scale raid on that city.

As the Bohlen attack force crossed the Rhine to safety, Bomber Command's operations for the night were only half complete. While the defenders' attention had been concentrated over central Germany, the raiding force of 166 Lancasters bound for Hemmingstedt ran in at low altitude, maintaining strict radio silence. Shortly before reaching their target the bombers rose above the radar horizon and began climbing to their attack altitude of 15,000ft. Each aircraft released large amounts of 'Window' to give the impression on radar that this was yet another feint attack. At 4.23 a.m. the attack on the refinery began, supported by jamming from a Fortress and a Liberator of No 100 Group. Because of the low-altitude approach and the clever use of 'Window', the German raid tracking organization failed to appreciate the strength of this force and the bombers were well on their way home before the first radar plots on 'weak formations' were reported in the target area. Night fighters were scrambled to engage the force but there were few interceptions and only one of the bombers was shot down.

During the two raids the oil refineries at Bohlen and Hemmingstedt were both hit hard and neither would resume production before the war ended. The night's action cost Bomber Command thirteen aircraft, including a Liberator and a Fortress of No 100 Group. Eight of the losses were attributed to attacks from night fighters and one to flak, two bombers were lost in a mid-air collision and the cause of the remaining two losses could not be established.

That night No 100 Group's Mosquitos had several skirmishes with German night fighters but only two of the latter were shot down—and both fell to Chapman and Stockley. The bombers' gunners claimed the destruction of two more enemy night fighters. German records indicate that the *Luftwaffe* lost seven night fighters that night, however. The fates of the other three aircraft will probably never be known but it is not difficult to speculate: a tired pilot, trying to land quickly on a dimly lit airfield patrolled by Mosquitos, might misjudge his approach and crash; a crew flying at low altitude to avoid being intercepted by a Mosquito might run into a hillside; a night fighter crew would switch off their IFF equipment to avoid betraying their position on 'Perfectos' and be shot down by 'friendly' anti-aircraft guns. Such losses, which were frequent, were the result of No 100 Group's efforts as surely as were those brought about by its night fighters. By this stage of the war the wide-ranging Mosquitos had became the bane of the German night fighter crews' existence.

When the war in Europe ended in May 1945, No 100 Group had honed its tactical stills to fine edge. Many of the electronic warfare techniques that it had pioneered for supporting bomber attacks would prove useful nearly more than forty-five years later, in the skies over Iraq.

NEW YEAR'S DAY PARTY

Operation 'Baseplate', the massed attack by Luftwaffe *fighters on Allied
airfields in France, Holland and Belgium, was originally intended to
neutralize the Allied air forces when the German Army launched its all-out
counter-offensive in the Ardennes in December 1944 aimed at recapturing the
important port of Antwerp. In the event the attack on the airfields took place a
couple of weeks after the offensive began, and its results were quite different
from those that the* Luftwaffe *planners had sought . . .*

DURING THE LATE autumn of 1944 German Army staff officers
laid elaborate plans for Operation 'Watch on the Rhine' (*'Wacht
am Rhein'*), a large-scale counter-offensive in the west. The
attacking force of 200,000 men, including seven *Panzer* divisions, was
to smash through the weakly held Ardennes sector of the US front and
thrust towards the port of Antwerp.

The *Luftwaffe* was assigned three main roles in the forthcoming
operation. First and most important, on the morning of the offensive it
was to mount large-scale strafing attacks on Allied forward airfields
with the intention of destroying as many planes as possible; second, it
was to seal off the battle area to prevent Allied aircraft from attacking
German troops and supply vehicles; and third, it was to fly ground-
attack missions in support of the advancing German troops.

Adolf Hitler had overseen the planning for 'Watch on the Rhine' and
Luftwaffe staff officers had been brought in only at a relatively late
stage when detailed planning was required for the air operations to
support the venture. The bulk of the aircraft to be sent against the
Allied forward airfields under Operation 'Baseplate' (*'Bodenplatte'*)
were to come from Reich air defence day-fighter units.

When he heard of the plan *Generalleutnant* Adolf Galland, the
Inspector of Fighters, was horrified and he said so in no uncertain
terms. In his view it represented a gross misuse of the air defence
units. He pointed out that ground attack was a specialized task and
one beyond many of his pilots, certainly the newer ones whose limited
training and flying experience had been orientated towards the air
defence role. Moreover, the majority of the fighter *Geschwader, Gruppe*
and *Staffel* leaders lacked the necessary training to lead formations
attacking ground targets from low altitude. Galland argued that such

an attack might achieve little and that there was a serious risk that the attackers themselves would suffer the heavier losses. The fighter ace's injunctions fell on deaf ears, however.

During December the fighter and ground-attack units assigned to Operation 'Baseplate' moved to airfields in western Germany. To preserve the secrecy of the move, the aircraft flew to their new bases at low altitude to remain below the cover of Allied radars, and they maintained strict radio silence. Each unit left behind radio operators at its previous airfield with orders to maintain the previous pattern of radio traffic by means of spoof transmissions.

On 14 December the commanders of the fighter and ground-attack *Gruppen* assigned to 'Baseplate' were summoned to the headquarters of *Jagdkorps II* at Altenkirchen, where they received a top-secret briefing on the planned operation. *Generalmajor* Dietrich Peltz, in command of the attacking force, informed his surprised audience that they were to prepare their units to take part in a series of near-simultaneous attacks on eighteen forward airfields used by the Allied air forces in France, Holland and Belgium. More than a thousand German aircraft were to take part in the operation, which was intended to suppress Allied air opposition during the opening stages of the ground offensive.

Two days later, before dawn on 16 December, a massive artillery bombardment heralded the opening of 'Watch on the Rhine'. Assault units moved forward and rapidly overwhelmed the defences at several points. The code-words that would have initiated Operation 'Baseplate' were not transmitted, however, and the units earmarked for the attack stayed on the ground. The land offensive was launched on that date on the basis of favourable weather forecasts, which predicted several days of low cloud and poor visibility over the battle area. For German Army commanders that offered a heaven-sent opportunity, for they knew that such weather would be more effective in keeping the Allied air forces 'off their backs' than anything the *Luftwaffe* could do.

During the two weeks following the Altenkirchen briefing the units assigned to 'Baseplate' flew air defence and ground-attack missions in support the offensive, though at low intensity om account of the poor weather. Since their commanders had heard nothing more about the attack on the Allied airfields, most of them assumed that the operation had been shelved. All the greater was their surprise when, on the afternoon of 31 December, the preliminary warning signal was issued informing them that the attack was to take place soon after first light the following morning.

That evening the German pilots were briefed on 'Baseplate' and on the part that each unit was to play in it. Then they were ordered to bed

MESSERSCHMITT Bf 109K-4

Role: Single-seat, high-altitude day fighter.
Powerplant: One Daimler Benz DB 605 ASCM 12-cylinder, liquid-cooled, inline engine developing 2,000hp at take-off.
Armament: One MK 108 30mm cannon firing through the propeller boss and two MG 151 15mm cannon mounted above the engine and synchonized to fire through the propeller disc.
Performance: Maximum speed 378mph at sea level, 452mph at 19,700ft; initial rate of climb 4,800ft/min.
Normal operational take-off weight: 7,475lb.
Dimensions: Span 32ft 8½in; length 29ft 0½in; wing area 174 sq ft.
Date of first production Bf 109K-4: October 1944.

to try to get a good night's sleep. The launching of the operation on New Year's Day had been prompted by a forecast of clear skies over Allied airfields the following morning. But if the previous evening's celebrations left many of those at the target airfields with hangovers that impaired their ability to fight back, so much the better.

Shortly after midnight on New Year's Day, four Arado Ar 234 bombers from *Kampfgeschwader 76* carried out the world's first-ever jet night bombing mission. The aircraft flew a circular route over Rotterdam and Antwerp, then they flew over Brussels and Liége, where they released their bombs at random. In fact the bombs were merely a diversion to allay Allied suspicions if the aircraft were tracked on radar; the true purpose of their mission was to provide a final check on weather conditions over the targets to confirm that 'Baseplate' could go ahead. The jet bomber pilots reported that the skies over Holland and Belgium were clear of cloud.

During the early-morning darkness the pilots assigned to 'Baseplate' were roused for their final briefings, and soon after dawn the attacking units took off and headed for their force rendezvous points. The air battles during the previous couple of weeks had taken their toll of the attack force and only 900 fighters and fighter-bombers, for the most part Messerschmitt Bf 109s and Focke Wulf FW 190s, took off for the attack. This total was somewhat less than that originally intended, but by any standard the attacking force was still a sizeable one.

Representative of the more successful attacks that morning was that mounted by *Jagdgeschwader 3* against the Royal Air Force base at Eindhoven in Holland. The airfield was home to eight squadrons of Typhoon fighter-bombers belonging to Nos 124 and 143 Wings and to three tactical reconnaissance squadrons of No 39 with Spitfires and Mustangs. Like many of the Allied forward airfields, this one was grossly overcrowded, with more than 150 aircraft dispersed around it plus visiting machines. RAF commanders would certainly have liked to have spread their units more thinly than this, but there was no choice: in the depths of the European winter only those airfields with an all-weather operating capability and hardened runways and taxiways could support sustained operations, and there were all too few of those usable in the forward areas.

The raiding force assigned to the attack on Eindhoven comprised *I*, *II* and *IV Gruppen* of *Jagdgeschwader 3*, the first two *Gruppen* operating Bf 109s, including some examples of the latest K-4 version, and *IV Gruppe* FW 190s. The three *Gruppen* assembled over the small town of Lippstadt and from there the eighteen four-plane sections, flying by sections in line astern, remained at low altitude, maintaining an almost straight track for Eindhoven 140 miles away. Thanks to the low-altitude approach and strict radio silence, the raiders arrived at their target with the advantage of surprise.

Jagdgeschwader 3 reached Eindhoven just as sixteen bomb-laden Typhoons of Nos 438 and 440 Squadrons were about to get airborne for an operation. The aircraft were bunched at the down-wind end of the runway and first two aircraft had begun to take off when the attackers commenced their strafing runs. The leader's Typhoon was badly shot up and the last act of its mortally wounded pilot was to cut the throttle and swing the aircraft off the runway. His wing-man continued with the take-off but was shot down and killed shortly after getting airborne. Seeing what was happening, the pilots of the remaining Typhoons unstrapped, struggled out of their cockpits and galloped clear of the planes before these too were shot up. The attacked units were both Canadian-manned, and the official Canadian history recorded that the Messerschmitts and Focke Wulfs 'attacked the field in a well-organised manner, being persistent and well-led'.

The strafing attacks on Eindhoven lasted about twenty minutes and resulted in the destruction of 30 Typhoons and reconnaissance aircraft and damage to fourteen more. Once they had overcome their initial surprise, the airfield's anti-aircraft gun defences fought back hard and, as a result of their efforts and skirmishes with Allied fighters during the withdrawal, *JG 3* lost about twenty aircraft. Similarly destructive attacks were mounted on the airfields at Brussels/Evère, Brussels/Melsbroek, St Denis-Westrem and Maldegem. At Metz-Frescaty, the most southerly airfield to be attacked, fighters of *Jagdgeschwader 53* caught the Thunderbolts of the 365th Fighter Group on the ground and destroyed about twenty of them.

Elsewhere the raiding forces were far less effective. For example, the US Army Air Force base at Asch in Belgium, home to the 352nd and 366th Fighter Groups, with Mustangs and Thunderbolts respectively, and three Royal Air Force Spitfire squadrons, came under attack from three *Gruppen* from *Jagdgeschwader 11*, two with FW 190s and one with Bf 109s. The raiders had the misfortune to arrive just after eight Thunderbolts had taken off and assembled into formation for a mission. On seeing the attackers, the American planes jettisoned their bombs and sped back to defend their base. *Leutnant* Georg Füdreder, a Bf 109 pilot with *II Gruppe*, recalled:

> Because of mist patches we took off later than planned, and I think this did a lot to impair the success of the mission. The *Gruppe* joined formation and set course for the target led by a couple of Ju 188 pathfinders . . . I did not notice whether any aircraft were hit by our own or by enemy flak on the way in, but as we neared the target I remember hearing on the radio that someone had been hit. Just short of the target we pulled up and fanned out to left and right to look over the airfield, then we went into our firing runs. I pulled up and went straight into my attack. My approach was too steep to engage the Thunderbolts on the east side of the airfield, so I aimed at four or five twin-engined planes in the north-west corner. I started a sharp 180-degree turn to go for the Thunderbolts on the east side, when tracer rounds streaked past me. At first I thought it was flak, then to my surprise I saw two Thunderbolts behind me. One was firing at me with everything but his aim was wild. I pulled sharply to port and his rounds passed astern of me. My pursuer and his No 2 gave up the chase and headed off west. I started after them, then broke away for a final run over the airfield, heading south. At this time I saw no other aircraft over or near the airfield. A pall of black smoke rose from the southern half of the airfield, coming from several burning aircraft. I made my firing run somewhat higher because of the smoke, but I still had to fly through the pall over the southern half of the airfield.

Colonel Harold Holt, Commander of the 366th, described the action from his viewpoint on the ground at Asch:

> The enemy was engaged immediately by a flight of eight of our T-bolts that had just taken off and assembled. Jettisoning their bombs, they attacked the enemy planes and kept them from hitting our pitifully unprotected planes on the ground. The entire air circus took place at tree-top level directly over the strip. Roaring engines, spitting machine guns and flaming planes going down to destruction had brought the war right to our door-step! The roadways around the strip were lined with spectators who dared leave fox-holes long enough to watch the show. But a plane turning into them soon caused a mad scatter for fox-holes. Slugs from a friendly plane could do as much damage as enemy lead . . .

The attackers destroyed seven Spitfires and some Dakota transports, and damaged several other aircraft; they also shot down four of the Mustangs that took off to engage them. But in achieving this *Jagdgeschwader 11* took a fearful beating from the defending fighters and anti-aircraft guns, losing nearly half of the sixty or so aircraft committed. Among the pilots lost were two skilful air defence aces, the *Geschwader* commander *Oberstleutnant* Günther Specht, credited with 32 victories, and *Hauptmann* Horst-Günther von Fassong, credited with 136. In some cases the German formations failed to locate their briefed targets altogether, and as a result the attacks on Volkel, Antwerp/Deurne and Le Culot airfields were all complete failures.

Several figures have been published concerning the number of Allied aircraft destroyed as a result of the *Luftwaffe* attack. The Royal Air Force official history states that 144 aircraft were destroyed and 84 damaged 'in the British area', but there is no knowing whether the figure refers only to aircraft destroyed on the ground or includes those that fell in the course of the numerous air combats that morning. US Army Air Forces records are even more vague and they indicate that about 30 USAAF planes were destroyed and several more suffered damage, though again it is not clear whether these figures include those lost in air-to-air combat. On the best available information, it appears that the Allies lost between 174 and 250 aircraft destroyed and between 84 and 120 damaged that morning. Considering the size of the engagement, Allied pilot losses were minimal, probably fewer than twenty in total. Within a couple of weeks of the attack all the Allied combat units would be restored to full strength, following the arrival of new or repaired aircraft from storage units.

From surviving German records we know that the attacks on the airfields cost the *Luftwaffe* 237 pilots killed, missing or captured, plus eighteen wounded. Among those lost were three *Geschwader* commanders, six *Gruppe* commanders and eleven *Staffel* commanders—experienced combat leaders who were impossible to replace. No official *Luftwaffe* figures survive for the number of aircraft that service lost in the operation. However, from the large number of pilots lost, it is reasonable to assume that these amounted to at least 300 planes or about a third of those that took part. Thus the *Luftwaffe* suffered greater losses in aircraft than those it inflicted, while its losses in pilots were more than ten times greater than those suffered by the Allied air forces. It was the greatest numerical loss, and also the highest proportional loss, ever to be suffered by a large force of attacking aircraft.

Operation 'Baseplate' was a catastrophe for the *Luftwaffe* fighter force, and several of the units involved never recovered from it. In destroying a commodity that the Allies possessed in abundance at that stage in the war, combat planes, the *Luftwaffe* expended the commodity it could least afford to lose, trained pilots. The action serves as yet another lesson about the dangers of sending airmen into action with inadequate training in their combat role. Military airfields that are well defended and at a high state of readiness are not easy targets. Even with a well-trained attack force, a low-level raid on such a target can be a risky business. Unless the operation is meticulously planned and skilfully executed, the raiding force is liable to lose more planes than it destroys. That was what happened during Operation 'Baseplate'.

12

TARGET HANOI

On 10 May 1972 US Air Force and Navy planes resumed their attacks on targets in North Vietnam, following a bombing pause lasting more than 3½ years. That day Air Force F-4 Phantoms delivered set-piece attacks on two targets in the vicinity of Hanoi, one of them with the newly developed 'smart' weapons fitted with electro-optical and laser-homing heads. This chapter describes these actions and the force package tactics employed by the attacking planes. An important innovation during this war, from the point of view of historians, was the installation of a tape recorders in combat aircraft; the action conversations reproduced in this chapter all come from this source.

DURING THE US Air Force raids on North Vietnam in the spring of 1972 the attacking planes flew from bases in Thailand. The F-4 Phantoms configured as bombers came from the 8th Tactical Fighter Wing at Ubon; the Phantoms configured for reconnaissance and air-to-air combat came from the 432nd Tactical Reconnaissance Wing at Udorn; and the EB-66 radar jamming planes, F-105F 'Wild Weasel' defence-suppression aircraft and EC-121 Warning Star airborne command and control planes came from the 388th Tactical Fighter Wing at Korat. Rescue and recovery planes to pick up crews who had been shot down came from the 56th Special Operations Wing at Nakhon Phanom, whilst the KC-135 tankers that supplied fuel for the high-speed jets before they entered enemy airspace and as they withdrew came from U Tapao and Don Muang.

On 10 May the Air Force's targets were the important Paul Doumer Bridge over the Red River, immediately to the east of Hanoi, and the nearby Yen Vien railway sorting yard. At 7.30 a.m. that morning the initial wave of seven KC-135 tankers (six aircraft required, plus an airborne reserve) began taking off from U Tapao each carrying 75 tons of fuel. As this was happening, an RF-4C Phantom conducted a weather reconnaissance of the Hanoi area; its coded radio report of clear skies over the targets allowed the preparations for the attacks to go ahead.

At the head of the raiding force entering enemy territory would be two four-plane units, 'Oyster' and 'Balter' Flights, with Phantoms armed for air-to-air combat. These took off from Udorn at 8.05 a.m. and headed north. Four more Phantom flights, configured in the same way, followed them into the air to escort different parts of the attack force over enemy territory. Last off from Udorn were a pair of RF-4Cs to carry out a post-strike reconnaissance of the targets.

From Korat, four EB-66 Destroyer radar-jamming planes took off to provide stand-off cover for the attack. Three flights of 'Wild Weasel' F-105G Thunderchiefs followed them, to provide defence-suppression cover for the attacks; because of the importance of this mission, each 'Wild Weasel' flight took off with five planes, including an airborne reserve that would turn back just short of enemy territory if all of the others were serviceable.

From Ubon, ten four-plane flights of Phantoms took off. Two carried chaff bombs to lay out a trail of the radar-reflective strips along the route to the targets. For the attack on the important Paul Doumer Bridge over the Red River, 'Goatee' Flight's planes were loaded with two 2,000lb electro-optically guided bombs. Those of 'Napkin', 'Biloxi' and 'Jingle' Flights were loaded with two 2,000lb laser-guided bombs. It was to be the first use of the two new 'smart' weapons against a target in North Vietnam. A further four flights of Phantoms, to attack the Yen Vien railway sorting yard, carried conventional, unguided 500lb bombs.

By 8.50 a.m. the entire armada was airborne and heading north. Of the total of more than 110 aircraft, no fewer than 88 were scheduled to penetrate enemy territory. Over northern Thailand the raiders refuelled from the six KC-135 tankers waiting at the rendezvous area, then headed into Laos. Already, however, the two flights that were to spearhead the attack had been weakened. Two planes from 'Balter' Flight had suffered technical problems and had to turn back. A Phantom of 'Oyster' Flight had suffered a radar failure, but its crew opted to continue the mission though with a reduced capability.

'Oyster' and 'Balter' Flights crossed the North Vietnamese border at 9.20 a.m. and headed for their patrol lines north-west of Hanoi. The plan was to establish a barrier patrol, with 'Oyster' Flight at low altitude and 'Balter' at 22,000ft some distance behind and in full view of enemy radars. Any MiGs moving against 'Balter' Flight were therefore likely to fly over 'Oyster' Flight waiting in ambush. Three minutes after crossing the frontier, the Phantom crews received warning from the EC-121 Warning Star radar-picket plane over Laos that enemy fighters were airborne. Similar warnings came from the cruiser USS *Chicago* acting as radar-control ship in the Gulf of Tonkin.

Initially the MiGs kept their distance from the incoming force, but at 9.42 the North Vietnamese controller finally ordered his fighters to go into action. A warning call from *Chicago* enabled Major Bob Lodge, 'Oyster' Flight Leader, to turn to meet the MiGs nose-on. That gave his

McDONNELL F-4E PHANTOM

Role: Two-seat, multi-role fighter.
Powerplant: Two General Electric J79-GE-17 turbojet engines each developing 17,900lb of thrust.
Armament: One M61A1 20mm cannon; (air-to-air) four AIM-7E Sparrow semi-active radar missiles and four AIM-9G Sidewinder infra-red homing missiles; (precision attack role during 10 May 1972 action) two 2,000lb electro-optically guided bombs or two 2,000lb laser-guided bombs; (normal attack role) nine 500lb bombs.
Performance: Maximum speed (clean) Mach 2.17 (1,430mph) at 36,000ft; maximum rate of climb (clean) 49,800ft/min.
Maximum operational take-off weight: 61,795lb.
Dimensions: Span 38ft 4in; length 63ft; wing area 530 sq ft.
Date of first production F-4E: June 1967.

flight a clear tactical advantage, for the Phantoms could engage the MiGs at long range with their Sparrow missiles while the enemy fighters with less advanced missiles had no effective means of shooting back.

Down at 2,000ft the Phantoms jettisoned their external tanks and accelerated to fighting speed. In an air combat, victory usually goes to the side which sees its opponent first. Lodge kept his force low to remain out of sight for as long as possible, while the four enemy fighters ran obliquely past his nose at 15,000ft. The two forces were not meeting exactly head-on, but even so their combined closing speed was tremendous—more than 1,000mph, or just over a mile every four seconds. Tape recorders in the fighters captured the words spoken as the MiGs appeared on their radars:

'Oyster 2 has contact!'
'Oyster 1 has a contact zero-five-zero [bearing] for fifteen [miles]'
'Oyster 3 is contact, Bob!'
'Right, we got 'em!'
'Oyster 1 on the nose, twelve miles, fifteen [degrees] high . . .'

The Phantoms were fitted with 'Combat Tree', a device that worked on the same principle as the Second World War British 'Perfectos' equipment (see Chapter 10). 'Combat Tree' transmitted signals to trigger the IFF transponders in the MiGs and picked up their coded reply signals. In the context of this mission the device was vitally important, for it provided proof that the planes seen on radar were hostile. And *that* meant that they could be engaged from long range with radar-homing missiles, without having to close to within visual range to confirm their identity. In the three 'Oyster' fighters with serviceable radars, the back-seater locked-on to an enemy plane and made ready the Sparrow missiles.

MIKOYAN-GUREVICH MiG-19 ('FARMER')

Role: Single-seat interceptor and air superiority fighter.
Powerplant: Two Tumansky RD-9B turbojets each developing 7,165lb of thrust with afterburning.
Armament: Three NR-30 30mm cannon.
Performance: Maximum speed (clean) Mach 1.4 (900mph) at 33,000ft.
Operational take-off weight (clean): 16,300lb.
Dimensions: Span 29ft 6½in; length 41ft 4in; wing area 269 sq ft.
Date of first production MiG-19: 1953.

Bob Lodge eased his fighter into a shallow climb preparatory to missile launch and the other Phantoms followed. When the leading MiG was in firing range, Lodge squeezed the trigger to launch his first Sparrow. Trailing smoke, the 450lb missile accelerated from the Phantom's 700mph to more than 2,000mph in 2.3 seconds. The motor then cut out and the weapon should have coasted on to the target, but instead it blew up in a cloud of smoke. Lodge squeezed the trigger a second time and another Sparrow streaked away from the fighter.

A few hundred yards to the right of Lodge, Lieutenant John Markle in 'Oyster 2' fired a pair of Sparrows. These gave similar results, as he later recalled:

> Our first missile apparently did not get rocket motor ignition. The second missile came off the aircraft and turned slightly right as it climbed. We continued to maintain position on 'Oyster 1' in an easy right turn, slightly nose-up. As I checked the missile's progress, the trail showed a slight left turn toward the radar target.

Captain Steve Ritchie in 'Oyster 3', 3,000yds to the left of Lodge, also launched a Sparrow but this too was a dud: the motor failed to ignite and it fell away from the fighter like a bomb. Thus, of the five Sparrows fired, three had failed to function properly. The two missiles

MIKOYAN-GUREVICH MiG-21bis ('FISHBED-J')

Role: Single-seat interceptor and air superiority fighter.
Powerplant: One Tumansky R-25 turbojet developing 16,500lb of thrust with afterburning.
Armament: One 23mm GSh-23 cannon; up to four K-13 infra-red homing, air-to-air missiles.
Performance: Maximum speed (clean) Mach 2 (1,320mph) at 36,000ft.
Maximum operational take-off weight: 22,000lb.
Dimensions: Span 23ft 6in; length 51ft 9in; wing area 247 sq ft.
Date of first production MiG-21bis: 1970.

that worked as advertised produced devastating results, however. The tape recorder in Markle's plane captured his reaction as he saw his missile explode beside the enemy plane:

'Oh right! . . . Now! . . . Good! . . . *Woooohooo!*' Then came a short pause to calm down, after which the pilot announced on the radio, 'Oyster 2's a hit!'

From the leading aircraft Lodge replied, 'I got one!'

Steve Eaves, Markle's back-seater, confirmed that he had seen the leader's kill: 'Roger, he's burning and he's going down one o'clock!'

Captain Roger Locher, Lodge's back-seater, saw a couple of small clouds suddenly appear as the two missiles detonated a long way in front of him. A few seconds later he caught sight of the two stricken North Vietnamese fighters, MiG-21s, tumbling out of the sky. One was cartwheeling and going down in a shallow dive, the other had part of a wing missing and was in a steep dive, rolling out of control.

The two MiG-21s that had survived the Sparrow onslaught flashed over the top of the Phantoms, as the latter were pulling into tight turns to the right to get behind their opponents for follow-up attacks. When Lodge rolled out he was only 200ft behind one of the MiGs, giving Locher in the rear seat a sight that he would never forget:

> We were in his jet wash. There he was, [after]burner plume sticking out, the shiniest airplane you've ever seen. He was going up in a chandelle to the right, we were right behind him.

The Phantom carried no gun and it was too close for a missile attack, so Lodge eased off the turn to open the range on the enemy fighter. It seemed only a matter of time before that MiG too was smashed out of the sky. Then, without warning, a pair of MiG-19s climbed steeply from below and gate-crashed the fight. Probably flown by Lieutenants Le Thanh Dao and Vu Van Hop of the 3rd Company, the newcomers slid into firing positions behind Lodge's Phantom as Markle bellowed a warning to his leader: 'OK, there's a Bandit . . . you got a Bandit in your 10 o'clock Bob, level!' The MiG-19s passed behind Lodge then closed in from his right.

Further warning calls followed from the other Phantoms: 'Bob, reverse right, reverse right Bob. Reverse right!'

Lodge's attention was focused on the MiG-21 in front of him, which by now had opened the range sufficiently to allow a close-in missile shot. Meanwhile, behind the Phantom, the leading MiG-19 opened fire with its cannon. The hefty 30mm rounds rapidly bridged the gap between the two machines and Markle's warnings to his leader took on a more strident note: 'He's firing—he's firing at you!'

In the Phantom under attack, events now followed each other in confusingly rapid succession. Roger Locher recalled:

> One or two seconds later—*wham!* We were hit. I looked up and saw the MiG [the MiG-21 in front of him] separating away. I thought we had mid-aired because that was exactly my interpretation of how a collision would feel. We both said 'Oh shit!', and my 'Oh shit!' was because the guy in front was getting away from us.

More 30mm shells slammed into the Phantom, and at last Locher realized what was happening. The fighter decelerated rapidly and he felt it yaw violently to the right. The whole of the fighter's rear fuselage was ablaze and as the flames ate their way forwards the heat began to roast the plastic of Locher's canopy, which turned an opaque orange. Smoke seeped into the cockpit.

Around the doomed Phantom the air battle continued. The MiG-21 that Lodge had been about to engage sped clear. The fourth plane from the original enemy formation was less fortunate, however. Steve Ritchie in 'Oyster 3' rolled out of his turn about a mile behind it, in a perfect firing position, and squeezed off two Sparrows in quick succession. Once again, the first missile failed to guide, but the other homed in perfectly. It detonated immediately below the Soviet-built fighter, pieces flew off it and the MiG quickly lost speed. As the Phantom swept past its victim, Chuck DeBellevue in the rear seat saw a black shape flash past his left, less than 100ft away—the enemy pilot. DeBellevue gave a jubilant shout on the radio: 'Oyster 3's a splash [enemy plane shot down]!'

DeBellevue's triumphant call was the last thing Roger Locher heard before he ejected from his Phantom. By then the burning plane was upside down and falling fast. Locher grabbed the firing handle between his legs with both hands, and pulled hard:

> We were under negative *g* at the time. My ass was off the seat; I was pinned against the top of the canopy. I saw the canopy go, then I went out under negative *g*. There was a lot of wind blast; I started to see again. Then *thwack!*—the parachute opened. And *Zoooom!*—past me went two MiG-19s.

The other members of 'Oyster' Flight watched in horror as the Phantom fell from the sky, hoping for a glimpse of one or more parachutes to indicate that there were survivors. They saw none. The fighter was upside-down when Locher ejected and probably the plume of smoke trailing behind the aircraft screened him from their view. It seems certain that Lodge was still in his cockpit when the fighter plunged into the ground.

Shaken by the loss of their leader, the survivors of 'Oyster' Flight kept up their speed as they returned to low altitude and made a rapid withdrawal from the area. The two aircraft of 'Balter' Flight also had a tussle with MiGs, though without loss on either side. Those skirmishes were not decisive, but for the raiding force now approaching Hanoi they had an important result: they kept the MiGs to the area north-west of city, enabling the main attackers to approach their targets without interference from enemy fighters.

Now the second phase of the action could begin, with the aim of blunting the cutting edge of Hanoi's SAM and anti-aircraft gun defences. Four EB-66E stand-off jamming aircraft, four 'Wild Weasel' F-105Gs and eight F-4 chaff-bombers were assigned to this task.

The first to make their presence felt were the EB-66s. Each plane carried a battery of eighteen radar-jammers and their role was similar to that of the target-support Liberators operated by No 100 Group in the Second World War (see Chapter 10). At 9.45 a.m. the aircraft arrived at their orbit positions just outside the reach of the Hanoi missile belt and, flying at 30,000ft, began jamming the enemy radars. As they did so four 'Wild Weasel' F-105Gs split into two pairs and ran into the defended area at low altitude. Their role was to strike at the radars of enemy surface-to-air missile sites as they came on the air. Each plane carried three anti-radar missiles—two short-range Shrikes and a long-range Standard ARM. One of those who flew on this type of operation, Major Don Kilgus, described the tactics used:

Once we had found an active site we would go into afterburner and increase speed to between 450 and 520 knots [520–600mph]. Speed gave us survivability and manoeuvre potential, because when we started pulling *g* the plane would slow down. The Thud [F-105] was a super plane but it was not the tightest turner in the world and you had to plan ahead if you wanted to make a violent manoeuvre. So we would light the burner, pull up and turn at the same time to establish the firing parameters for the missile. Needles showed where the Shrike was looking and gave an indication of range plus or minus about 20 per cent. If a loft attack was necessary, for maximum range we would launch in a 30-degree climb from 15,000 feet. Or if we were close to the site we would pitch over and push the missile down his throat.

The 'Wild Weasel' aircraft were themselves likely to come under missile attack, and this was no mission for the faint-hearted.

As the F-105s played their deadly game of hide-and-seek with the SAM batteries, the next phase of the action opened. Eight Phantoms ran in towards Hanoi from the south-west at 26,000ft, each carrying nine chaff bombs. The role of these planes was similar to that of the 'Window Spoof' forces in the Second World War (again, as described in Chapter 10). At 9.47 a.m. these aircraft entered the Hanoi SAM-defended zone and each released a single chaff bomb. After a short fall the casings split open and each disgorged millions of metallized strips each thinner than a human hair. At 15-second intervals along the route to the enemy capital, each plane released a further chaff bomb.

During the run-in the Phantoms flew in the so-called 'jamming pod formation', with two lots of four aircraft flying in line abreast with 2,000ft horizontal separation and stepped up to one side with 600ft vertical separation between adjacent planes. This formation offered a high degree of protection against the SA-2, the only long-range missile system then used by the North Vietnamese. Each Phantom carried a jamming pod under the fuselage, and the noise-jamming from the four-plane formation produced a wedge of overlapping strobes on the enemy gun and SAM control radars. The SAM operators could see the incoming formation, but they could not pick out the individual planes accurately to engage them.

That, at least, was the theory. It worked only if the crews held their positions in formation. As the Phantoms closed on the enemy capital the crews watched the clouds of orange or white smoke on the ground as the enemy missiles blasted away from their launchers. During its boost phase each missile left a smoke trail, but when this ended there was nothing to see until then missile itself hove into view. Captain William Byrns recalled:

A SAM came for us and someone yelled 'Look out!' I turned my head and my reaction was to pull back on the stick. That was not the normal reaction—I should have gone down. But I believe God took my hand and made me go the other way. The missile went underneath my plane, underneath the F-4 across the way and exploded on the far side of him. If I had gone down it would have hit us and we would probably not have got out

Several of the crews experienced similar scares, and a few planes suffered a shaking as missiles detonated within a few hundred feet. Yet the protective cocoon of jamming conferred a high degree of safety: only one chaff bomber was hit by missile splinters, and these caused little damage.

Having released the last of their chaff bombs, the two flights sped away from the target area. Behind them they left more than seventy clouds of chaff that now spread out to form a corridor two miles wide, more than a mile deep and eighteen miles long. At the end of that corridor lay the Paul Doumer Bridge.

Five minutes behind the chaff bombers, the first of the main attack formations entered the Hanoi missile zone. Flying through the corridor of chaff laid to assist it, the Paul Doumer Bridge attack force headed for the target flying at 620mph at 13,000ft. 'Goatee', 'Napkin', 'Biloxi' and 'Jingle' Flights, each with four Phantoms in jamming-pod formation, followed each other at two-mile intervals.

Flying ahead of the Phantoms and far below them, a fresh team of four 'Wild Weasel' F-105Gs fanned out in pairs, looking for active SAM sites. Yet despite this harassment and the radar jamming from the Phantoms and their supporting EB-66s, the defending batteries lay on an impressive display of wrath. Holding position in a jamming-pod formation under SAM attack has been likened to the first time one snuffs out a candle with one's fingers—it was an unnatural act and it required courage to overcome one's basic instincts. Captain Lynn High commented:

> We had to sit in formation and grit our teeth when the SAMs came through the formation. It took nerves of steel to watch a SAM come straight at you, even though you knew that in all probability it would not hit you and if it detonated it would detonate too soon or too late. I watched about six SAMs do exactly that.

Meanwhile the leading attack flight, 'Goatee', commenced its bombing run on the Paul Doumer Bridge with electro-optical guided bombs (EOGBs). These weapons homed in on the image contrast of the target against its background, and the Phantoms were to attack the bridge broadside-on from the south. In each plane the back-seater operated a small hand-held controller to position the bridge under the sighting reticle on his TV screen and pressed a button to lock the target image into the first bomb. He then switched to the second bomb and repeated the process. Colonel Carl Miller, the Flight Leader, pushed his plane into a 30-degree descent and the other three planes followed. At 12,000ft the Phantoms released their bombs in a salvo.

During the recent war in the Persian Gulf the newest types of EOGBs demonstrated an impressive degree of accuracy. Earlier weapons of this type were considerably less effective, however, and on this occasion they performed miserably. The exasperated Miller watched his bombs go their separate ways:

> One made a 90-degree turn and went for downtown Hanoi, I think it impacted near the train station. I don't know where the other went. The

EOGB was a launch-and-leave weapon, they were supposed to stay locked-on after release. But they didn't.

After releasing their bombs, the aircraft of 'Goatee' Flight turned west and engaged their afterburners to get clear of the defended area as rapidly as possible. As this was happening, 'Biloxi' 'Jingle' and 'Napkin' Flights sped towards a point immediately to the east of the bridge, ready to attack along its length with their laser-guided bombs (LGBs). As he was about to turn in to attack, Major Bill Driggers glanced to his left to observe the result of 'Goatee' Flight's attack with EOGBs. He had expected to see them burst around the bridge and demolish parts of the structure, but the reality was quite different:

> As we rolled in to attack the bridge I saw big waterspouts rising from the Red River. The first EOGBs fell short; those that made it to the bridge went through the gaps between the pylons. I saw two, maybe three, of the bombs explode in the water on the other side.

None of the EOGBs hit the target—a profoundly disappointing result during the first operational use of these expensive weapons over North Vietnam.

Two by two, the Phantoms of 'Napkin' Flight pulled into their 45-degree attack dives. The 37mm and 57mm anti-aircraft guns around the target opened up a powerful defensive fire and colourful lines of moving tracer, punctuated by stationary puffs from exploding shells, criss-crossed the sky over the eastern side of Hanoi. At 12,000ft each Phantom let go its bombs and pulled out of its dive. In the leading plane in each pair, the back-seater operated a small control stick to hold the laser-designator on the required aiming-point. A laser seeker head in the nose of each bomb steered the weapon to the point thus marked. The first salvo of four bombs exploded against the bridge or in the river beneath it, hurling smoke, spray and debris hundreds of feet into the air.

At the head of 'Biloxi' Flight, the next to attack, Captain Lynn High noticed that the enemy anti-aircraft gunners seemed to be aiming at the wrong part of the sky:

> The Vietnamese gunners obviously expected us to release from a lower altitude: they coned their fire on a point 7,500 feet to 9,000 feet above the target. It looked like an Indian tepee sitting over downtown Hanoi. But we released our bombs at a higher altitude—we kept out of it.

The Flight's eight 2,000-pounders threw up further columns of debris, smoke and spray around the bridge.

'Jingle' Flight bombed last, and Captain Mike Van Wagenen piloted the final aircraft to attack the bridge:

> There was so much going on, it was impossible to comprehend everything. The human mind cannot take that many inputs so it rules a lot of them. The radio seemed to go quiet, the radar warning gear went quiet, everything appeared to go quiet as I tracked the Doumer Bridge underneath my sighting pipper. We just stopped thinking about the other things going on around us. My back-seater was calling off the altitudes: 15 ... 14 ... 13 [thousand feet] ... The pipper was tracking up the bridge, I had the parameters like I wanted to see them and released both bombs.

Van Wagenen hauled on the stick and watched the horizon sink rapidly past his windscreen as the *g* forces asserted themselves and pushed him hard into his seat:

> As we came off the target it was like plugging in the stereo: slowly one's senses came back and one could hear the radar warning gear, the radio transmissions, everything else. The human computer was working again. I jinked hard left and right, picked up Mike [Captain Mike Messett, his element leader] and joined up on him. Then I rolled back to the right to see where my bombs had gone. It appeared all four, Mike's and mine, had hit the first span on the east side of the river. I took one more look to see if the span was standing but I couldn't tell, there was a lot of smoke around.

As Van Wagenen left the bridge none of the spans had dropped, despite the fact that several of the laser-guided bombs had scored direct hits on the structure and caused severe damage. Two spans at the eastern end had broken apart, however, and the bridge was impassable to wheeled vehicles.

As the Phantoms sped out of the target area some of them had fleeting brushes with MiGs. Lieutenant Rick Bates recalled:

> As we came off the target we passed a Thud [F-105] followed by a MiG followed by a Thud. Then I saw a MiG-21 that looked as if it was trying to turn on us. But we were going so goddarn fast he had no chance ... Those three or four minutes was [*sic*] absolute and total chaos as far as I was concerned; my pulse rate was going at about eight million a minute ...

As the Phantoms of the bridge attack force left the target, the raiding force heading for the Yen Vien rail yard began running through the chaff corridor laid earlier. The sixteen F-4 bombers followed the same route as the bridge attack force and Major Kelly Irving was surprised at the ease with which he could follow the line of chaff through the defended area:

I was impressed at how well it showed up on my air intercept radar. That was how we made sure we were positioned in it. That was a godsend—we drove up that thing like it was a highway.

The bombers ran towards their target at 15,000ft, pulled up to 20,000ft and swung into echelon right as they peeled into their 45-degree attack dives. Captain Jim Shaw, in the leading flight, recalled:

We followed the other three down the chute and Bud Pratt [his pilot] pickled [released] the bombs. When we pulled off I looked back, and saw somebody's bombs do a pretty good job across the south chokepoint. While in the target area we tried to change something—heading, altitude or speed—every ten seconds to defeat the radar-aimed fire.

This attention to detail proved necessary, for as Shaw left the target his flight suddenly came under heavy fire:

Lead got away with it, No 2 flew through some of it, No 3 could not avoid it. We broke left and came very sharply back to the right. I got an eyeful of all the standard colours of smoke puffs. The larger the calibre the darker the smoke: white puffs were 23mm, light grey puffs were 3 mm, grey were 57mm and black puffs were 85mm. Beforehand every flight leader briefs that he will fly a wide arc coming off the target, so those behind can cut off the turn and join up for mutual support. But when they were being shot at, very few leaders do it to the degree their wing-men would like. We went out scalded-ass fast and it took a while to get the flight back in order. Everybody had a distinct interest in getting away from the people they had just been nasty to!

The attacks on both targets were now complete, but there was a further chore that had to be completed. North of Hanoi a pair of RF-4C reconnaissance Phantoms moved into position to photograph the targets for damage assessment. The aircraft accelerated to 750mph, keeping just below the speed of sound to retain a measure of manoeuvrability, and sped towards their objectives at between 4,000 and 6,000ft, continually varying their altitude to give the enemy gunners as difficult a target as possible. Major Sid Rogers led the pair, with Captain Don Pickard flying as wing-man behind and about 1,000ft to the right of his leader. The latter recalled:

After we passed the rail yard we got everything in the world shot at us. We started jinking and as we approached Hanoi there was a trail of black puffs from bursting shells behind Sid. I said to my back-seater, Chuck Irwin, 'Good God, look at that stuff behind lead!' Chuck replied, 'Its a good thing you can't see the stuff behind us . . . !'

Moments later Pickard noticed a MiG-17 about 500yds behind and to the left, trying to get into a firing position. But the two Phantoms soon left the slower fighter far behind. South of Hanoi a SAM battery loosed off at Pickard's aircraft. The pilot saw nothing of the missile until it detonated and the Phantom bucked under its blast:

I didn't see the SAM but I saw a whole bunch of red things, like tracer rounds but fanning out, come past my nose [the hot splinters from the warhead]. I ducked; it looked like we were going to hit them.

Miraculously, all the warhead splinters missed the plane.

It was 10.14 a.m. and now all the bombers, the reconnaissance planes and the 'Wild Weasel' and jamming-support aircraft were heading away from the target. Four flights of Phantoms covered the withdrawal: one patrolled north-west of Hanoi, one was to the south-south-west, one was to the south-south-east and one was astride the withdrawal route near the Laotian border.

The earlier activity by MiGs had tapered off and now there was little sign of the defending fighters. Tempted by this inactivity, one flight moved west of Hanoi at 8,000ft trying to lure North Vietnamese fighters into battle. The stratagem succeeded only too well. Suddenly a MiG-19 zoomed into a firing position behind one of the Phantoms and delivered a snap attack. Exploding 30mm shells tore away chunks from the left wing and the fighter rolled into a dive and plunged into the ground. There were no survivors. The remaining Phantoms curved vengefully after their assailant and one launched missiles from extreme range, but the North Vietnamese pilot knew his business and dived away and disappeared as suddenly as he had come.

By 11.15 a.m. the whole of the raiding force was back on the ground. The 8th Tactical Fighter Wing at Ubon had sent out forty F-4s to lay the chaff corridor and deliver the attacks on the Paul Doumer Bridge and the Yen Vien railway yard; as a testimony to the effectiveness of the 'jamming-pod formation', all the planes had passed through the thickest part of the defences and all had returned, though one had suffered minor damage. The 388th Tactical Fighter Wing at Korat sent twelve F-105s, four EB-66s and an EC-121 to support the operation; all these returned safely too. The 432nd Tactical Reconnaissance Wing at

Udorn had sent 28 F-4s and three RF-4Cs; its two F-4s shot down by MiG-19s were the only planes lost during the mission. Three North Vietnamese MiG-21 fighters had been shot down, all of them by 'Oyster' Flight during the initial encounter.

On the following day Phantoms delivered a second attack with laser-guided bombs on the Paul Doumer Bridge, during which they concentrated their weapons on the damaged eastern end of the structure. After further hits, the disconnected span toppled into the Red River.

Just over three weeks later there was an interesting sequel to the action. It will be remembered that Captain Roger Locher had ejected from his blazing Phantom shortly before it crashed into the ground. Although deep in enemy territory he avoided capture, living off any edible vegetation he could find. On June 1 Locher finally made radio contact with US aircraft, and the following day a large-scale rescue operation retrieved him. When he was picked up by a 'Jolly Green Giant' helicopter he had lost 30lb in weight and he was weak from starvation. For an air-crew survivor to remain at liberty, unassisted, for 23 days deep in enemy home territory and initiate a successful rescue was a record for the Vietnam War and it ranks with the most successful combat evasion episodes in history. The rescue also set a record for those who retrieved him, for it took the helicopters deeper into North Vietnam than any other such mission during the conflict.

13

THE FIRST 'BLACK BUCK'

Following the Argentine capture of the Falklands Islands early in April 1982, the British government dispatched a large naval force to repossess the islands. At the airfield at Waddington in Lincolnshire the three resident Avro Vulcan bomber units, Nos 44, 50 and 101 Squadrons, received orders to prepare a number of aircraft and crews to fly extended-range missions over the South Atlantic. The code-name for the Operation was 'Black Buck' . . .

WHEN ROYAL Air Force planners began to examine the feasibility of using the elderly Vulcan, the only available long-range bomber type, to support the operation to retake the Falklands, it became clear that this would be no easy option. The distance from Port Stanley, the capital, to the nearest available airfield on Ascension Island was 3,886 statute miles—roughly equal to the distance from London to Karachi in Pakistan. The 7,700-mile round trip was a good deal further than that flown on any previous operational bombing mission in history. Although in its time the delta-winged bomber had been modified to refuel in flight, the system had been out of use for more than a decade and scarcely any of the generation of crews then flying the bomber had received training in its use.

Even if the bomber's air-to-air refuelling system could be brought back into use, to get a Vulcan loaded with bombs to the Falklands and back again would entail an enormous supporting operation by tankers. It would require no fewer than ten Victor tanker sorties to supply the bomber and the tanker that were to accompany it to the distant south, and a further tanker would have to meet the bomber and provide fuel for the final part of the return flight. The big question was: would the damage that a single Vulcan could inflict on the enemy possibly be worth the expense and effort required to mount the operation?

While the decision whether to use the Vulcan was being pondered, work began to prepare half a dozen of the bombers for the operation, 'just in case'. The first move was to refurbish the planes' in-flight refuelling receiver systems and, as these became ready, the selected Vulcan crews began training to take fuel from Victor tankers. As the planning of the operation reached the detailed stage, a series of problems became apparent. One of the most daunting was the inadequacy of the Vulcan's navigation system for the unforeseen mission. The aircraft was equipped to operate in areas where there were plenty of land features on which its H2S bombing radar could obtain fixes, and the radar and navigational computer system were quite unsuitable for use over the featureless wastes of the South Atlantic, where fixing points were few and far between. Yet, perhaps short of fuel, the bomber would need to be able to make a rapid and accurate rendezvous with the tanker that was to replenish its tanks during the return flight. To make up for this deficiency, the Vulcans and the Victor tankers were modified to carry the Carousel inertial navigation equipment as part of the programme to prepare them for the operation.

With most complex military operations, if there is time it is usual to mount a rehearsal beforehand. In the case of the 'Black Buck' operations there was time available, but no attempt was made to make a full-scale test the validity of the in-flight refuelling operation. As one of the pilots later explained,

It would have been as much trouble to run a rehearsal as to fly the mission, so it was decided to fly the mission. If the problems had become too great, we would have broken off the mission and called it the rehearsal . . .

As April drew to a close it became clear that there was unlikely to be any diplomatic solution to the crisis between Great Britain and Argentina. The matter would have to be settled by force of arms. In the final week of the month two of the specially prepared Vulcans flew to Wideawake, each loaded with twenty-one 1,000lb bombs. The crews had already been briefed on their target for the initial attack, the runway at Port Stanley airfield. Soon after they arrived on Ascension Island they learned that the mission was to take place during the small hours of 1 May. One Vulcan was assigned to fly the mission; the other would take off with it, to act as a reserve in case the primary machine went unserviceable.

From 10.50 p.m. Ascension time (7.50 p.m. Port Stanley time) in the evening of 30 April the two Vulcan bombers and the eleven supporting Victor tankers (including one airborne reserve) thundered into the air from Wideawake at one-minute intervals. As the aircraft headed south, their anti-collision lights blinking, the wisdom of the planners in providing airborne reserves was quickly borne out. Soon after take-off it was discovered that the primary bombing Vulcan was unable to pressurize its cabin; and that the fuel hose unit of one of the Victor tankers was unserviceable. Both planes therefore abandoned the

AVRO VULCAN B.2

Role: Five-seat medium bomber (a sixth crew member, an air-to-air refuelling expert, was carried during the 'Black Buck' missions).
Powerplant: Four Rolls-Royce (Bristol) Olympus 301 turbojet engines each developing 20,000lb of thrust.
Armament: (Conventional attack role) Normal operational bomb load 21,000lb. During some 'Black Buck' missions the aircraft carried two or four Shrike anti-radar missiles.
Performance: Cruising speed Mach 0.93 (707mph) at 45,000ft; combat radius of action without refuelling 1,725 miles; service ceiling 58,000ft.
Normal operational take-off weight: 200,000lb.
Dimensions: Span 111ft; length (inc. refuelling probe) 105ft 6in; wing area 3,964 sq ft.
Date of first production Vulcan B.2: Spring 1960.

mission and returned to Wideawake. The force, now lacking any margin in capacity if there were a further major failure, continued on its way south.

The Captain of the reserve Vulcan was Flight Lieutenant Martin Withers. When he received the news that the primary aircraft had aborted the mission, those on board the aircraft remember that there was a long and pensive silence on the intercom. Then Withers commented, 'Looks like we've got a job of work, fellers . . .' No further discussion was necessary, for the reserve crew had briefed as assiduously for the mission as had the one now forced to abandon it.

For one and three-quarter hours after take-off the gaggle of big jets headed south, then, at a point about 840 miles from Ascension, four of the Victors passed their spare fuel to four others and turned back. Another Victor passed fuel to the Vulcan.

Even at this early stage, a problem manifested itself that was to cause increasing difficulties as the operation progressed. Holding loose formation, the Vulcan and its attendant Victors flew at a compromise cruising speed that was the optimum for neither machine; moreover, their cruising altitude of 31,000ft was chosen because it was the highest at which fuel could be transferred between them, and it was somewhat lower than the height for optimum fuel consumption. The result was that both types of aircraft consumed fuel slightly faster than expected. The four Victors that had given up their fuel at the first transfer had to dip deeply into their own reserves in order to pass on the required amount to those continuing south; this would give rise to another problem a few hours later.

The second fuel transfer took place two and a half hours after take-off, about 1,150 miles south of Ascension. A Victor topped up the Vulcan's fuel tanks, then turned back. Soon afterwards two Victors passed fuel to the three Victors continuing south, then they too turned back.

Four hours after take-off there were tense scenes at Wideawake, as the four Victors that had been the first to give up their spare fuel arrived almost simultaneously at the airfield. All of the planes were low on fuel. The single east–west runway runs between rocky outcrops, and it could be entered or left only at its western end. As luck would have it, the wind was from the east, which meant that each Victor's landing run took it to the end of the runway opposite from the exit. In normal circumstances each plane would have landed, then stop, turn round on the runway and taxi to the exit point and be clear of the surface before the next aircraft landed. But now the circumstances were not normal, and had the Victors followed the usual procedure the last couple of aircraft in the queue might have run out of fuel completely before they could put down.

The alternative was not ideal, but it was the only course of action that was feasible in the circumstances. The first Victor touched down, ran to the far end of the runway and stopped; the second aircraft landed and pulled up close behind the first; and the third tanker landed and pulled up close behind the second. When Squadron Leader Martin Todd made his approach, at the controls of the fourth Victor, the stage was set for the aeronautical equivalent of a motorway pile-up. Had there been any misjudgement on the part of the pilot, or a relatively minor technical failure in his aircraft, the Royal Air Force stood to lose one quarter of its available tanker force in the South Atlantic area. Furthermore, a couple of these machines were earmarked to take fuel to the aircraft that in a few hours would be returning from the distant south: if anything prevented that, the entire force heading south would have had to be recalled.

Todd placed his Victor firmly on to the runway and pulled the handle to stream the plane's huge braking parachute. He felt a reassuring push as his body was pressed against the seat straps, and

---

HANDLEY PAGE VICTOR K.2

Role: Five-seat in-flight refuelling tanker aircraft.
Powerplant: Four Rolls-Royce Conway 201 turbofan engines each developing 20,600lb of thrust.
Armament: Nil.
Fuel capacity: 127,000lb.
Performance: Cruising speed Mach 0.92 (700mph) at 45,000ft.
Normal operational take-off weight: 240,000lb.
Dimensions: 117ft; length 114ft 11in; wing area 2,597 sq ft.
Date of first Victor K.2: March 1972 (rebuilt B.2R reconnaissance aircraft).

the aircraft decelerated rapidly. In front of him sat the three Victors, invisible in the darkness but for the insistent blinking of their anti-collision lights. Later he commented:

> There were the other three at the end of the runway, waiting for us to stop. If our brakes had failed or anything—Christ, I hate to think of it . . .

But there was no failure. Todd pulled up well short of the other three aircraft, turned his Victor through a semi-circle and taxied to the runway exit. In relieved procession the other three tankers followed him.

As that drama was being played out, the third transfer of fuel began 1,900 miles south of Ascension. Flight Lieutenant Alan Skelton passed all his spare fuel to two of the other Victors, then turned back for Ascension. Soon afterwards, however, he discovered that his aircraft had developed a fuel leak. The quantity being lost was not large and in normal circumstances it would not have mattered, but he was a long way from Ascension and he had bitten deeply into his own fuel reserves in order to the pass as much as possible to the other aircraft. One of Skelton's crew called Ascension and asked that a tanker be sent to meet them on the way back to the island.

The force heading south was now down to two Victors and the single Vulcan. Five and a half hours and 2,700 miles after take-off, there was a further transfer of fuel. And again there was an unexpected problem. Squadron Leader Bob Tuxford, captain of one of the Victors, explained:

> There is an unwritten rule in air-to-air refuelling, a variation of Sod's Law, which says, 'If you're going to find any really bumpy weather, it will be right at the point where you have to do your tanking.' Now that proved to be the case and the 'really bumpy weather' duly appeared as a violent tropical storm at exactly the point where the final transfer of fuel between the Victors was planned to take place.

From the cockpit of the Vulcan Martin Withers observed the shadowy outline of a Victor a few hundred yards to his left trying to take fuel from another:

> It was dreadfully turbulent, we were in and out of the cloud tops, there was a lot of electrical activity with St Elmo's Fire dancing around the cockpit. The Victor was trying to refuel in that—he was having enormous problems. We could see the two aircraft bucking around, with the refuelling hose and basket going up and down about 20 feet.

Eventually, after some superb flying, Flight Lieutenant Steve Biglands succeeded in pushing his refuelling probe into the basket streamed behind Tuxford's aircraft. The fuel transfer began, but the triumph was short-lived. Suddenly Biglands gave a terse radio call to say that his refuelling probe had broken. That put the entire operation in jeopardy, for the tanker could not take on any more fuel during the flight and there could be no question of it accompanying the Vulcan to the far south. The only alternative was for the two remaining Victors to exchange roles, with Biglands giving up his spare fuel to Tuxford so that the latter could continue south with the Vulcan.

By the end of the transfer the final pair of aircraft were more than 3,000 miles south of Ascension and the Vulcan was just over an hour away from its target. In the Victor there was an earnest discussion on whether it would be possible to continue the mission. Bob Tuxford continued:

> We were considerably lower on fuel than we should have been. Now we had a decision to make: either to go on, give the Vulcan the fuel it needed to make its attack, and prejudice our own position because if we didn't pick up some more fuel on the way back we would have to ditch; or turn back at that stage while we both had sufficient fuel to get back to Ascension. I was the Captain of the aircraft and I had to make the decision, but I asked my crew what they thought. One by one they came back and said, 'We have to go on with the mission.'

Because of the need to keep radio traffic to an absolute minimum, the Vulcan crew had no inkling of the problems facing those aboard the tanker. The two aircraft linked up for the final transfer of fuel before the target, at a point about 400 miles north-east of Port Stanley. The transfer went ahead normally until, when the Vulcan's tanks were about 6,000lb short of full, Martin Withers was disconcerted to see the red indicator lights on the underside of the Victor flash on to indicate that the fuel transfer was complete. Withers broke radio silence with a brief request for more fuel, but Tuxford told him curtly that there could be no more. Later the refuelling captain commented:

> Not being familiar with the tanking game, not knowing how far I had stretched myself to put him where he was, all he knew was that he wanted a certain amount of fuel. If only he had realized how much discussion had already taken place in my aeroplane, about how far we could afford to stretch ourselves to get him there . . .

Having taken their decision, the Victor crew had now to live with its stark terms: unless they could summon another tanker to pass more fuel to them, their aircraft would inevitably crash into the sea about 400 miles south of Ascension. Moreover, because of the necessity that the Argentine forces on the Falklands should have no inkling of what was afoot, the crew could not use their high-frequency radio to inform base of their predicament until the Vulcan had completed its attack. Although Withers had less fuel than planned, the Vulcan had sufficient for the operation and he knew that a Victor tanker, plus a reserve, were scheduled to meet him during the return journey to top up his tanks.

When the bomber reached a point 290 miles from Port Stanley, Withers eased back the throttles and the Vulcan began a slow descent to keep it below the horizon of enemy early-warning radars on the Falklands. At 2,000ft he levelled off and continued towards the target. Flight Lieutenant Bob Wright, the radar operator, switched on his radar transmitter for a few seconds and observed returns from the top of Mount Usborne, the highest point on East Falkland. That brought heartening confirmation that, throughout the long over-water flight, the Carousel inertial navigation equipment had kept the bomber almost exactly on the planned track.

Shortly after 4 a.m. (local time), at a point 46 miles from the target, Withers pushed forward the throttles to bring the Vulcan's four Olympus engines to maximum thrust. As speed built up he eased the bomber into a steep climb to bring it to its briefed attack altitude of 10,000ft. Once there, the pilot levelled out and let his speed build up to 400mph before easing back on the throttles to hold that speed. Meanwhile the radar operator again turned on his transmitter and the crew settled into the bombing run. The attack was aimed at the mid-point of the runway, with the aircraft running in at an angle of 30 degrees its length. Bearing in mind the Vulcan's 1950s-vintage aiming system, that gave the greatest chance of scoring at least one hit on the runway (an attack down the length of the runway would have produced several hits if everything went perfectly, but a slight error in line would have caused all the bombs to miss).

During the bombing run Withers saw nothing of the target in the darkness below him. His job was to follow as accurately and as smoothly as possible the left/right steering signals on the display in front of him generated by the attack computer. Later he recalled:

> It was a smooth night, everything was steady, the steering signals were steady and the range was coming down nicely. All of the switching had been made, and ten miles from the target we opened the bomb doors. I was expecting flak and perhaps missiles to come up but nothing happened. The AEO [Air Electronics Officer] didn't say anything about the defences and I didn't ask—I left that side of things entirely to him. I was concentrating entirely on flying the aircraft.

Bombers Against the *Tirpitz*

A<small>BOVE</small>: The 42,900-ton battleship *Tirpitz*, showing her powerful main armament of eight 15in guns. During the final action on 12 November 1944 she used the four weapons in the two forward turrets to defend herself.

B<small>ELOW</small>: A close-up view of the 'Tallboy' bomb. Hits with two of these weapons blew away a large area of plating in the port side of *Tirpitz*, and the resultant inrush of water caused the battleship to capsize. (IWM)

Confound and Destroy

LEFT: Air Vice-Marshal Edward Addison, commander of No 100 Group, pictured at his headquarters at Bylaugh Hall near Dereham, Norfolk. (Addison)

ABOVE: A B-24 Liberator radar-jamming aircraft of No 233 Squadron.

BELOW: A 'Jostle' high-powered radio-jamming transmitter, seen on its specially modified transporter vehicle. This jammer was carried by No 100 Group's Fortress and Liberator aircraft.

Above: A Halifax elint aircraft of No 192 Squadron.

Below: Right lower: Mosquito night fighters of No 85 Squadron, fitted with special equipment to enable them to home-in on German night fighters from long distances.

Right upper: A Junkers Ju 88G night fighter. (Obert)

Right lower: A Messerschmitt Bf 110 night fighter shot down by a Mosquito of No 85 Squadron.

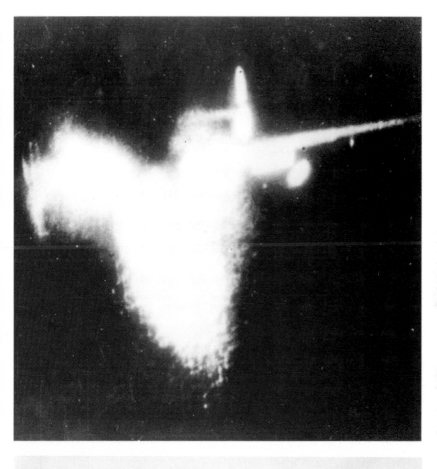

New Year's Day Party

BELOW: *General* Adolf Galland, Inspector of Fighters, argued strongly against committing a large part of the home defence fighter force for Operation 'Baseplate', but to no avail.

RIGHT: *Oberstleutnant* Günther Specht (nearest camera), commander of *Jagdgeschwader 11*, and *Oberstleutnant* Johan Kogler (wearing peaked cap), commander of *Jagdgeschwader 6*, were both shot down during Operation 'Baseplate'. Specht was killed and Kogler was taken prisoner. At that stage of the war the *Luftwaffe* fighter force could ill afford to lose such experienced commanders. (Kogler)

RIGHT LOWER: A Messerschmitt Bf 109G of *III Gruppe* of *Jagdgeschwader 53*, one of the units that attacked the US airfield at Metz-Frescaty during Operation 'Baseplate'. (Via Schliephake)

Target Hanoi

Above: Colonel Carl Miller led the attack by F-4 Phantoms on the Paul Doumer Bridge near Hanoi on 10 May 1972. (Miller)

Right upper: An F-4D Phantom of the 8th Tactical Fighter Wing that took part in the attack on 10 May, showing the asymmetric loading of stores on planes fitted with laser-marking pods: from right to left—2,000lb laser-guided bomb, Pave Knife laser-marking pod, centreline 600 US gallon fuel tank, 2,000lb LGB, 370 US gallon fuel tank. (USAF)

Right lower: A North Vietnamese SA-2 'Guideline' missile on its launcher. Thanks to clever tactics and jamming of the missile-guidance radars, although several of these weapons were fired during the action on 10 May they shot down no aircraft.

Above: The Paul Doumer road/rail bridge over the Red River at Hanoi, pictured after the war. The light-coloured areas of the structure show sections that had been damaged during the various bombing attacks and were afterwards replaced. (Lye)

Below: North Vietnamese MiG-19 'Farmer' fighters shot down two Phantoms during combats on 10 May. These fighters proved extremely agile, and in each case their pilots delivered snap attacks on US fighters coming up from below their victims.

The First 'Black Buck'
Right: A view from the cockpit of a Vulcan as it moves towards a Victor B(K).2 tanker to take fuel.

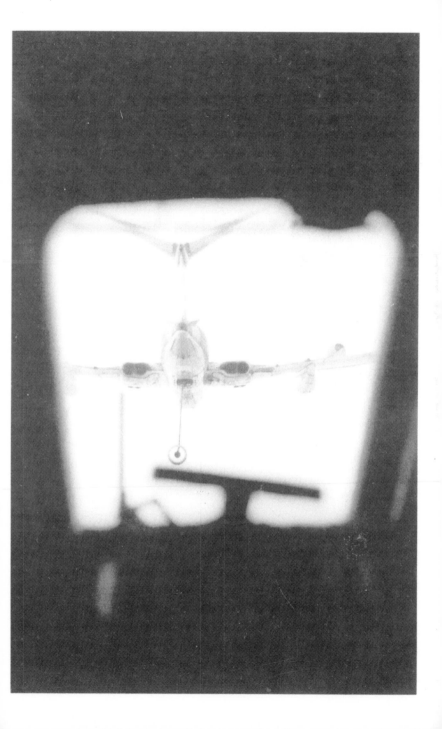

Above and below: The Vulcan coming in to land at Wideawake following the mission to the Falkland Islands on May 1982. (MoD)

Right upper: A Victor tanker landing at Wideawake airfield, Ascension Island, after a refuelling mission.

Right lower: Flight Lieutenant Martin Withers, pictured after his return from the attack on Port Stanley airfield. (MoD)

Countdown to 'Desert Storm'
Aʙᴏᴠᴇ: McDonnell Douglas F-15C Eagle fighters provided cover during the initial attack and they shot down four Iraqi aircraft. (USAF)

Bᴇʟᴏᴡ: Boeing E-3 Sentry AWACS aircraft kept watch over the Coalition air attack forces and directed the escorting fighters into position to attack Iraqi planes. (USAF)

ABOVE: Royal Air Force Panavia Tornado GR.1 aircraft carrying a pair of JP.233 airfield-denial weapons containers under the fuselage. (Panavia)

BELOW: A General Dynamics F-111F of the 48th Tactical Fighter Wing, pictured immediately before taking off for a mission, with the ground crews removing the safety pins from the four 2,000lb laser-guided bombs. (USAF)

Tornado Spyplanes Go To War
Above: An infra-red linescan picture taken from a Tornado GR.1A, showing Iraqi trenches to the west of Wadi al Khirr during the early morning darkness on 26 January 1991. In the picture the vehicle tracks, although they are more or less straight, follow an elongated 'S' because of the distortions produced by the horizon-to-horizon coverage of the system. The distortions do not detract from the military value of the picture, however. (MoD)

In fact the Argentine defenders were preparing to engage the plane bearing down on the airfield. Flight Lieutenant Hugh Prior, the Vulcan's Air Electronics Officer, picked up signals from gun-control radar trying to lock-on to the bomber; he flicked a switch to turn on his radar jammer and the signals ceased.

When the Vulcan reached the bomb-release point, the attack computer triggered the bomb-release mechanism and the twenty-one bombs were dropped at ¼-second intervals. As the last bomb left the aircraft, Withers ordered the bomb doors to be closed and he pushed open his throttles. Then he hauled the bomber into a steep turn, to get himself clear of the defended area as quickly as possible. After a fall lasting 20 seconds, the bombs exploded in a neat line across the airfield.

As the Vulcan turned away, Flying Officer Pete Taylor, the co-pilot, glanced to his right and made out the street lights in the town of Port Stanley. Then, much closer, he saw a series of flashes in quick succession below the thin layer of cloud that covered the airfield. As the last bomb exploded, the darkness returned and those in the Vulcan felt rather than heard the crump of the distant explosions.

Nobody in Port Stanley that night will ever forget the sound of those detonations. Artist Tony Chater and his wife Ann were in bed at their home in the centre of town:

> I was half awake at the time and the whole house shook. It was as though there had been an earthquake, then we heard the *boom boom boom boom* of the bombs going off, very muffled. Shortly afterwards I just made out the sound of an aircraft in the distance.

To the Falklanders, the sound of that opening attack provided an enormous fillip to morale. Chater continued:

> There was terrific jubilation in Stanley. From then on we really felt very confident that the British forces were going to come to our rescue.

The Vulcan was well out of range and climbing to altitude when the Argentine gun positions situated around the airfield finally came to life. Their noisy but ineffectual display of defiance continued for several minutes, then one by one the guns fell silent.

In the cabin of the bomber there were no feelings of jubilation to match those of the citizens of Port Stanley. The nervous exertions of the previous eight hours had drained the crew of much of their emotional energy, and later Withers summed up the mood:

> After the attack the crew were very quiet, rather sad. We had just started a shooting war. It had all been rather cold-blooded, creeping in

there at 4.30 in the morning to drop bombs on the place. But we had a job to do and we thought that job worth doing.

The bombs cut a swathe of destruction across the middle of the airfield. The first bomb landed on the runway close to its mid-point, penetrated the surface and detonated to cause a large hole with considerable 'heave' around the lip. The second bomb clipped the southern edge of the runway, causing similar damage. One of the other bombs in the stick detonated between the airfield's sole repair hangar and a Pucará attack plane parked nearby, causing damage to both. Yet another bomb blew out the windows in the control tower and gave the building a severe shaking. Three Argentine military personnel were killed and several injured. Considering the age and the known limitations of the Vulcan's attack system, the raid was as effective as might reasonably have been expected.

As the Vulcan continued its away, Hugh Prior broadcast the code-word 'Superfuse' to announce that the attack had been carried out and appeared to have been successful. That signal was the cue for Bob Tuxford's Victor to break radio silence to inform base that the aircraft had insufficient fuel to reach Ascension, urgently requesting that a tanker meet it on the way back. The crew were out on a limb—just a minor failure away from disaster. Later Tuxford commented:

It was a long, dry journey back. We discussed a lot of things, including the practical aspects of bailing out of a Victor into the sea—you would not try to ditch it, the aircraft was the wrong shape. We had our radar on to see if there were any ships in the area, but in fact there was none in the right place.

In the event the Tuxford made a successful rendezvous with the tanker scrambled to meet him. So did Alan Skelton who, it will be remembered, had suffered a fuel leak early in the operation.

Just over four hours after the attack, the Vulcan reached its planned refuelling point off the coast of Brazil. The sun was high in the sky as Martin Withers caught sight of the white underbelly of a Victor swinging into position in front of him. The tanker levelled out with the hose trailing invitingly behind. It was, he later commented, 'the most beautiful sight in the world.'

Withers advanced slowly on the Victor and pushed his refuelling probe into the basket on the end of the hose. Initially the precious fuel flowed smoothly into the bomber's tanks, but as the pressure built up it began to spill from the connection. The translucent liquid gushed over the plane's windscreen and even with the high-speed wipers going the pilots could make out only the blurred outline of the aircraft in front. The visibility ahead was rather like that from a vehicle being driven through a car wash.

Had this been a normal training sortie Withers would have throttled back, broken contact, then moved forward again to insert the probe properly into the refuelling basket. But his bomber was low on fuel, and there was a chance that the refuelling probe or the basket had suffered damage. If he broke contact now he could not be certain that he could regain it. Although some fuel was being lost, most of it was flowing into the bomber's tanks. For each minute that Withers could maintain the precarious contact, his bomber took on a further ton of fuel.

Then help came from an unexpected quarter. Bob Wright, the Vulcan's navigator, was standing on the ladder between the pilots' seats watching the operation. As the fuel gushed over the canopy he noticed that, almost level with his eyes, at the base of the centre windscreen, the airflow was keeping a narrow strip of glass clear of fuel. Through this he could see the tanker clearly, and was able to give the pilots a running commentary on relative position of aircraft in front to assist them to hold the contact.

Withers took ten minutes to take on the fuel he needed. Then he throttled back to break contact with the tanker, and as the probe withdrew from the basket a valve shut off the supply of fuel to the hose. In an instant the airflow cleared the fuel away from the Vulcan's windscreen and suddenly there were sunshine and blue skies outside the bomber's cockpit. Withers felt as if a huge burden had been lifted from his shoulders:

After that fuel was on board, the other four hours back to Ascension were a bit of a bore. Only then was the tension off and we knew we were going to make it. Those four hours seemed to last for ever.

The Vulcan landed at Wideawake after just over sixteen hours after it had taken off. Later Martin Withers received the Distinguished Flying Cross for the leadership he displayed during the attack, while Bob Tuxford received the Air Force Cross for the selfless manner in which he and his crew had supported it.

So ended the first 'Black Buck' mission by a Vulcan. The operation stretched the capabilities of the bomber, the Victor tankers and all the crews involved to the very limit. In retrospect, it is clear that the effort expended in the operation was out of all proportion to the physical damage that it inflicted on Port Stanley airfield. Yet, as is often the case in aerial warfare, the raid on Port Stanley airfield had a psychological effect on the enemy that was also out of proportion to the physical damage it caused. The attack demonstrated to the Argentine Air Force High Command that the Vulcans had the capability to strike at targets on the Argentine mainland at any time. On the following day that service's only specialized interceptor squadron, *Gruppo 8* with Mirage III fighters, was withdrawn from Rio Gallegos in the south of the country, to where it had moved to support operations over the Falklands. To meet the new threat the unit transferred to Comodoro Rivadavia much further north, and, apart from a single skirmish near the end of the conflict, *Gruppo 8* played no further part in the fighting. In effect, the Argentine Air Force had conceded defeat in the battle for air superiority over the Falklands. From then on the Royal Navy's Sea Harriers were allowed to hunt down the enemy fighter-bombers and attack planes without having to worry about themselves being preyed upon by enemy fighters. That was the 'bottom line' result of that first 'Black Buck' mission, and *that* justified the enormous effort that had been expended.

14

COUNTDOWN TO 'DESERT STORM'

At the end of November 1990 the United Nations Security Council passed Resolution 678, which stated that unless Iraqi forces withdrew from Kuwait by 15 January the following year member states would be permitted to employ 'all necessary means' to dislodge them. President Sadam Hussein believed that his armed forces were strong enough to hold the territory he had seized and decided to call what he considered to be the bluff of the powers aligned against him. As the deadline approached it became clear that no amount of diplomacy could resolve the crisis. What the Iraqi dictator had grandiloquently termed 'The Mother of Battles' was about to begin. The action that took place on the first night of the conflict provided a vivid insight into the nature of aerial warfare during the final decade of the twentieth century.

D URING THE LATTER half of 1990 the Coalition of nations arrayed against Iraq moved large numbers of troops and a huge contingent of aircraft into Saudi Arabia and the surrounding states. By the following January preparations were well advanced for Operation 'Desert Storm', the large-scale aerial onslaught against targets in Iraq. As the Security Council deadline passed, the Coalition command staffs laid final plans for the attack to begin with a massive air strike during the small hours of 17 January.

Throughout the period of tension US and Royal Saudi Air Force Boeing E-3 Sentry AWACS aircraft had flown round-the-clock patrols to keep watch on the movements of Iraqi aircraft and to provide advanced warning of a possible attack on Saudi Arabia itself. Ready to go into action to meet such a threat were standing patrols of the most modern fighter aircraft available to the western air forces—F-14 Tomcats, F-15 Eagles, F/A-18 Hornets, Tornado F.3s and Mirage 2000s. After the deadline passed, the AWACS planes and the fighters maintained the same operating patterns as before, to keep the Iraqi defence forces in ignorance of what was in store for them.

The main attack on the Iraqi air defence system, and on miliary targets in that country and Kuwait, was to commence at 3.00 a.m. Baghdad time on the morning of the 17 January. This was designated 'H-Hour' for the operation. The first aircraft to get airborne specifically to attack Iraq took off from Barksdale, Louisiana, at 6.35 a.m. Central Standard Time on the morning of the 16th (H-Hour minus 11 hours 25 minutes). Seven B-52 Stratofortresses of the 2nd Bomb Wing, each loaded with five AGM-86 cruise missiles, set out for a strike on targets in northern Iraq more than 7,000 miles away. Just over seven hours later, at H minus 4 hours, a further twenty B-52s began taking off from their island base at Diego Garcia far away in the Indian Ocean.

The next to take off, at H minus 2 hours, were the slowest of the attacking aircraft scheduled to go into action that night—eight AH-64 Apache attack helicopters of Task Force 'Normandy' of the US Army's 101st Airborne Division. With them went two Air Force MH-53J 'Pave Low' heavy-lift helicopters fitted with special electronic equipment that were to serve as navigational 'mother ships' for the operation. The helicopter force left Al Jouf airfield in the northern part of Saudi Arabia at 1 a.m.

The mission of the attack helicopters was to knock out two strat-egically placed air-defence radars west of Baghdad, thus opening up a corridor through which high-speed jet attack planes could pass unseen on the way to their targets. The helicopters were due to hit the radars shortly before the jets entered their areas of cover. The requirement to achieve surprise and the highest possible chance of destroying the radars, and the need for an immediate assessment of damage after the attack, had led to the decision to employ attack helicopters rather than fixed-winged aircraft. In addition to a large external fuel tank, each helicopter carried eight Hellfire laser-guided missiles, a pod with nineteen 70mm unguided rockets and a built-in armament of one 30mm cannon in a turret mounted under the nose. Their crews observing the terrain around and the 'mother ships' in front of them through night-vision goggles, the blacked-out Apaches hugged the desert floor as they flew north at a stately 120mph. Soon afterwards the first of 160 tanker aircraft—KC-135s, KC-10s, KA-6As, Victors and VC-10s—began taking off and moved into position to provide refuelling support for the initial waves of attacking aircraft.

At 1.31 a.m., H minus 1 hour 29 minutes, the first Tomahawk cruise missile roared away from the deck of the cruiser USS *San Jacinto* at the northern end of the Red Sea. Once it was clear of its launcher, the missile's wings unfolded and it descended to low altitude and made for the pin-point target programmed into its attack computer. Soon afterwards the cruiser *Bunker Hill*, then the battleships *Wisconsin* and *Missouri* in the Persian Gulf, joined in the bombardment. That night the four ships loosed off a total of 52 cruise missiles.

Shortly after the first of the sea-launched missiles had been sent on its way, the seven B-52s that had flown direct from the United States arrived at their missile-launching points over the north of Saudi Arabia. The heavy bombers launched a total of 35 cruise missiles, which were programmed to hit eight communications, air defence and airfield targets in the Mosul area. As the cruise missiles dropped to low altitude and sped across Iraq, the big bombers turned around and began the long haul back to Barksdale. Their total time airborne would be 34 hours 20 minutes, making this the longest bombing mission ever and exceeding the previous longest, those by the Vulcans to the Falkland Islands (described in the previous chapter), by a considerable margin.

At 2.15 a.m., H minus 45, the first of the bombers began to take off— F-117As, F-111s, F-15Es, A-6s and Tornados, with F-14s, F-15s and F-18s 'riding shotgun' to protect them from enemy fighters. F-4G Phantoms, A-7 Corsairs, EF-111 Ravens and EA-6B Prowlers accompanied the raiders to suppress the defences in the target areas. Wing Commander John Broadbent described his take off from Muharraq airfield, Bahrain, at the head of a force of eight RAF Tornados:

We taxied out under radio silence, and took off singly on a green light from tower. Then we climbed in trail to 10,000 feet and joined up with the Victor tankers that had taken off ahead of us. We took on fuel going along the route to the target.

As the cruise missiles flew unswervingly over their pre-programmed routes, the F-117A Stealth Fighters moved unseen into position to strike at their assigned targets and the teams of attack planes began their low-altitude penetrations into Iraqi territory, a quite different and more visible air operation unfolded high above the Iraqi SAM defences. Timed to begin shortly before the attack planes entered the missile defended zones, this involved large numbers of unmanned decoys carrying echo-enhancement equipment to give them a radar signature similar to that from a full-sized aircraft. Launched from ground sites located in the north of Saudi Arabia, 38 Northrop BQM-74 pilotless drones crossed the border into Iraq flying at medium altitudes and at speeds around 575mph. Powered by a 180lb thrust turbojet, the 5ft-span vehicle had been designed as an expendable target drone and that was now to be its task, literally. Some of the drones headed for the missile-defended zones around Basrah in the east of the country while others made for the H2 and H3 airfield complexes in the west. The drones entered the defended areas in groups of three or four, creating the illusion of combat planes flying in tactical formation.

At the same time another type of lure, the US Navy's Tactical Air Launched Decoy (TALD), was used bring to life the missile and gun defences in the Baghdad area. Each US Navy A-6 Intruder taking part in the operation carried eight of these lightweight unpowered decoys folded up under its wings. The planes released the decoys from altitudes around 20,000ft, and once each TALD was clear of its launch aircraft its wings unfolded and it began its silent glide towards the defended area.

The incoming decoys made tantalizingly easy targets, and the Iraqi ground batteries launched salvo after salvo of expensive and irreplaceable SAMs at them. Well-trained SAM crews might not have been fooled by such a simple stratagem, but those who manned the missile sites that night were not in that category. The first independent confirmation that the war had begun came at 2.37 a.m. Baghdad time, when CNN television broadcast the now-famous pictures showing tracer rounds arcing across the night sky over the Iraqi capital punctuated by the occasional explosions from shells or missiles. That was 21 minutes before H-hour and long before the real attack on targets in the capital was scheduled to begin, and it is almost certain that the television cameras had in fact filmed the Iraqi reaction to the TALD 'attack'.

That night the ground defences enjoyed something of a 'turkey shoot', causing great slaughter among the American decoys. Unfortunately for several of the defenders, however, these particular 'turkeys' had a more sinister purpose. Once the Iraqi SAM batteries had been drawn into a full-scale defensive action, the operation entered its second phase. Flying behind the decoys, and somewhat lower, were several F-4G 'Wild Weasel' and F/A-18 aircraft carrying AGM-88 HARMs (High speed Anti-Radiation Missiles). Set to home on the transmissions from the enemy fire-control radars, the missiles were launched in large numbers. For one brief period during the onslaught there were no fewer than two hundred HARMs in flight and closing rapidly on the enemy emitters.

The elaborately planned 'spoof and punch' attack destroyed or damaged the fire-control radars at several SAM sites, putting the associated missile battery out of action until the radar could be repaired or replaced. At other sites the battery had fired off all its immediate-use missiles, and it would take several minutes to hoist the reload missiles on to their launchers and make them ready for action; the sites would still be engaged in this task when the Coalition attack forces swept past them on the way to their targets.

Meanwhile the Apache helicopters of Task Force 'Normandy' arrived in firing positions on the two radars they were to attack (the latter had been left alone long enough to report the approaching decoys). The helicopter attack commenced at 2.38 a.m., H minus 22. Using infra-red night vision systems to laser-mark the targets, the helicopters fired their Hellcat missiles from ranges of about three miles, then closed to a

PANAVIA TORNADO GR.1

Role: Two-seat, swing-wing attack and reconnaissance aircraft.
Powerplant: Two Turbo-Union RB.199 turbofasn engines each rated at 15,000lb thrust with afterburners.
Armament: (Carried by all Tornado GR.1s during the first might of the War) Two JP.233 airfield-denial weapons containers weighing a total of 10,300lb; two Mauser 27mm cannon; two AIM-9L Sidewinder infra-red homing missiles for self-defence.
Performance: Maximum speed (at low altitude with full weapons load) 680mph, (at high altitude, clean) Mach 2.2 (1,450mph) plus.
Maximum gross take-off weight: 60,000lb.
Dimensions: Span (wings fully forward) 45ft 7in, (wings fully swept) 28ft 2in; length 54ft 9½in; wing area (wings fully forward) 323 sq ft.
Date of first production Tornado GR.1: June 1979.

mile and a half to deliver their unguided rockets. The various parts of targets were demolished in strict order of priority: first the electrical generators, then the communications facilities, then, finally, the radars themselves. In less than two minutes the Apaches fired 27 Hellfire missiles, about a hundred 70mm rockets and some 4,000 rounds of 30mm ammunition, which reduced both radar stations to smoking ruins. Their task completed, the helicopters turned round and retraced their flight paths to friendly territory.

By now the teams of attack planes were well into their low-altitude penetrations in Iraqi territory. Initially there was no visible reaction from the defences and several attacking crews felt an air of unreality about the proceedings. Lieutenant Dave Giachetti, Weapons System Officer in one of eight F-111Fs of the 48th Tactical Fighter Wing on their way to attack chemical weapons storage bunkers at Ad Diwaniyah near Baghdad, remembered:

I thought it was kinda eerie, because outside everything was so calm and so quiet. We went in at low level on TFR. In the built-up areas everyone had their lights on; the street lights were on. On the way in we flew parallel to a road for some time; there were cars moving with their lights on. We were at flying at 400 feet at 540 knots [620mph] towards our target and I thought, man, they don't even know we're coming!

Flight Lieutenant 'Moose' Poole, navigator in one of the RAF Tornados heading for Al Taqaddum airfield, was another of those who had difficulty coming to terms with the fact that he really was flying an operational mission that night:

There was no moon and it was very, very dark. The ground was as flat as a witch's tit and there was little apparent movement of the aircraft.

There were no visual clues outside; the only light was from my instruments. I found myself becoming detached from reality—it was just like being in the simulator flying a war sortie. I had to tell myself that this was no simulator sortie—it was for real.

Each crew caught up with the realities of the situation in its own way. Wing Commander Jerry Witts, leading four Tornados heading for an attack on Mudaysis airfield, found his moment of truth a few minutes after he crossed into Iraq when there was a chilling reminder that war could be injurious to the health:

We had just crossed the border, lights off and sneaky beaky, when on the RHWR [radar homing and warning receiver] we got what looked like Fulcrum [MiG-29 radar] spoke at 2 o'clock. If there was anything that worried me at that stage of the war, it was the look-down/shoot-down capability of the Fulcrum. We did an evasive turn and the spoke duly trotted around the aeroplane, which is exactly what one would expect if it was moving round on to our rear. I thought, Jesus—I've only been at war for five minutes . . .

Soon afterwards the 'Fulcrum radar spoke' disappeared. Almost certainly it had been a false alarm.

By now the initial salvos of cruise missiles from the ships and the B-52s were in the final stages of their approach on their targets. Only a relatively small proportion of these missiles carried high-explosive warheads for the direct attack on the enemy air defence system; most of the Tomahawks carried warloads that, while not in themselves lethal, were designed to have a devastating effect on the Iraqi defence system. Their payload consisted of a large number of small spools, each measuring ½in by ¾in, wound with a long length of electrical-conducting carbon-fibre resembling thin electrical wire. In making a detailed analysis of the Iraqi air defence system, Coalition Intelligence officers had discovered a major weakness that could be exploited: the

---

LOCKHEED F-117A

Role: Single-seat, precision-attack stealth aircraft.
Powerplant: Two General Electric F404-GE-F1D2 turbofans without afterburners, each rated at 10,800lb thrust.
Armament: Typical operational bomb load two 2,000lb LGBs or four 500lb LGBs. No defensive armament.
Performance: Classified, but maximum speed high subsonic.
Maximum gross take-off weight: 52,500lb.
Dimensions: Span 43ft 4in; length 65ft 11in; wing area 912.7 sq ft.
Date of first production F-117A: Spring 1982.

---

computers and the operations centres took their power from the national electricity supply grid. If these could be deprived of mains electrical power, the working of the entire control system would be halted until back-up supplies could be brought on line—and vital information on the unfolding air battle would be lost when the computers 'crashed'. Several of the cruise missiles were programmed to fly low over power stations and electrical power lines, spewing out the spools as they did so. Once it was free in the airflow, each spool unwound rapidly to lay out a serpentine length of carbon fibre that drifted slowly to the ground. If the fibre fell across a high-voltage electrical conductor, it produced a massive short circuit that caused severe local damage to the electrical transmission system. In contrast to the sophisticated electronic countermeasures systems also being employed that night, this was a simple but extremely effective 'electrical countermeasure'.

In a further move to prepare the path for other attack forces, at 2.51 a.m. (H minus 9) an F-117A stealth fighter dropped laser-guided bombs on an important Iraqi air defence operations centre in the south-west of the country. The first attack by a manned aircraft on Baghdad itself took place at 3 a.m., H-hour, when one of the stealth fighters attacked a communications centre in the city. That night F-117As attacked 34 targets associated with various aspects of the Iraqi air defence system.

As the attack forces penetrated progressively deeper into Iraq, the US fighters escorting the raiders had their first encounters with enemy planes. Captain Steve Tate of the 71st Tactical Fighter Squadron was leading a flight of four F-15C Eagles, providing cover for strike packages moving into the Baghdad area. From its patrol line over Saudi Arabia, the E-3 AWACS aircraft controlling Tate's flight reported that an unidentified aircraft was apparently closing on the No 3 aircraft in the flight. Tate later recalled:

> My Number 3 had just turned south, and I was heading north-east on a different pattern. I don't know if the bogey [unidentified aircraft] was chasing him, but I locked him up [on radar], confirmed he was hostile and fired a missile.

Launched from a range of twelve miles, Tate's AIM-7 Sparrow sped towards the rapidly closing Iraqi plane—which by then had been identified as a Mirage F.1—and shortly afterwards the American pilot saw a fireball as the weapon impacted:

> When the airplane blew up, the whole sky lit up. It continued to burn all the way to the ground and then blew up into a thousand pieces.

McDONNELL DOUGLAS F-15C EAGLE

Role: Single-seat air superiority fighter.
Powerplant: Two Pratt & Whitney F100-PW-220 turbofans each rated at 23,450lb thrust with afterburners.
Armament: On normal operations the armament carried was four AIM-7M Sparrow semi-active radar homing missiles and four AIM-9M Sidewinder infra-red homing missiles, and an internally mounted M61A1 20mm cannon.
Performance: Maximum speed (at high altitude) Mach 2.5 (1,900mph) plus, (at low altitude) 921mph.
Normal operational take-off weight: 44,630lb.
Dimensions: Span 42ft 9¾in; length 63ft 9in; wing area 608 sq ft.
Date of first production F-15C: February 1979.

Following this initial success, marauding F-15s had several fleeting radar contacts with Iraqi planes. Owing to the presence of Coalition attack forces in their vicinity, however, the AWACS controller refused to give clearance for the American fighters to attack with their long-range missiles: the Sparrow could be lethal to friend or foe alike, and it was vitally important to ensure that any aircraft engaged was clear of Allied planes. Captain Larry Pitts of the 33rd Tactical Fighter Wing, one of the F-15 pilots airborne that night, recalled:

> Our radar scopes were filled with friendlies—60 to 80 of them! Night conditions combined with bad weather made it difficult to fire missiles even if the F-15s acquired targets. There were just too many friendlies out there.

Then one of the MiG-29s moved clear of Allied planes long enough for it to be singled out for attack, and Captain Jon Kelk of the 33rd TFW destroyed it with a Sparrow. Five minutes later Captain Robert Grater of the same unit scored the first double kill of the war, with the

MIKOYAN-GUREVICH MiG-29 ('FULCRUM')

Role: Single-seat interceptor fighter.
Powerplant: Two Isotov ID-33 turbofan engines each rated at 19,000lb of thrust with afterburners.
Armament: Up to six air-to-air missiles on underwing pylons, usually a mixture of the R-27 medium-range, semi-active radar homing missile (AA-10 'Alamo') and the R-73 (AA-11) or R-60 (AA-8) infra-red homing dogfight missile; one 30mm cannon.
Performance: Maximum speed (at high altitude) Mach 2.3 (1,520mph).
Maximum gross take-off weight: 60,000lb.
Dimensions: Span 37ft 3¼in; length 56ft 10in; wing area 378.9 sq ft.
Date of first production Mig-29: 1982.

destruction of two Mirage F.1s in quick succession, also using Sparrows. Over Vietnam the performance of the semi-active radar homing missile had sometimes been disappointing, but over Iraq the improved version then in use was proving to be a highly effective weapon.

Leading the formation of Tornados making for Al Asad, Wing Commander Ian Travers Smith watched one of the Iraqi fighters go down:

> We had not seen any activity at all—no AAA. The first sign that I was really at war was when an aircraft suddenly burst into flames in our 11 o'clock. There was scuddy cloud at medium altitude and there was this great fireball falling from the sky, with bits coming off.

A few Iraqi fighters managed to close on Coalition attack planes that night, but the latter were carrying air-to-air missiles for self-protection and they were well able to look after themselves. A twelve-plane force of F-15E Strike Eagles of the 4th Fighter Wing, running in at low level to attack a 'Scud' missile launching site near H-2 airfield, had a couple of MiG-29s approach it. It appears that, in attempting to slide into an attacking position behind an F-15E flying at high speed about 100ft above the ground, one of the Iraqi pilots misjudged his height. The MiG smashed into the ground and exploded without a shot being fired on either side.

The disruption to the Iraqi air defence control and reporting system that night prevented it from keeping track of the large number of attacking forces sweeping over the country. As a result, at several of the targets there was little or no warning of the raiders' approach until their bombs detonated. The simultaneous attack on the airfields began at H plus 1 hour, 4 a.m.

The 48th Tactical Fighter Wing, the most powerful US Air Force night precision-attack unit, sent 53 F-111Fs in forces of between four and six aircraft to hit a dozen separate targets that night. These included the major airfields at Balad and Jalibah in Iraq, and those at Ali Al Salem and Al Jaber in Kuwait, where the hardened aircraft shelters were thought to contain 'Scud' missiles. F-111Fs also attacked chemical weapons storage bunkers at H-3 airfield, Salman Pak and Ad Diwaniyah.

Six F-111Fs hit the airfield at Balad, with two aircraft launching GBU-15 electro-optical guided bombs at the maintenance complex while the other four dropped large numbers of area-denial mines at each end of the runways and among the aircraft shelters. Further west, another six F-111Fs attacked the chemical weapons storage bunkers at H-3 airfield with LGBs. As they pulled away from the target, four Royal Saudi Air Force Tornados attacked the runways. Supporting that attack was a defence-suppression force from the carrier USS *John F. Kennedy* in the Red Sea comprising three EA-6B jamming aircraft, ten A-7s carrying HARMs, and four F-14 fighters.

At the eastern end of the war zone Lieutenant-Colonel Tommy Crawford led six F-111Fs against Ali Al Salem airfield in Kuwait. His target was the hardened aircraft shelters beside the airfield, thought to house 'Scud' missiles. On the way in the bombers flew over Iraqi Army units, which put up a disconcerting amount of tracer into the sky above them. Crawford recalled:

> We made a low-level attack because of the SAM threat—there were several SA-6 launchers in the area. Our intention was to run in at 1,000 feet until SA-6 signals on the RHWR forced us down. But there was so much AAA—I couldn't believe how much there was—we crossed the border at 2,000 feet and that was where we stayed until we delivered the bombs. It seemed like every 50 yards there was a guy with a gun who was shooting up at us—it was the damnedest 4th of July show you ever saw. As we approached the border it looked like a solid wall of fire, but you have no perception of depth so it looked a lot worse than it really was. Once we had crossed the border it seemed the flak opened up in front of us as we flew along, and then it seemed as if it was worse to the sides and behind than it was in front.

The F-111Fs tossed their GBU-24 laser-guided bombs at the shelters from a range of about four miles, then curved away from the target as the WSO held the laser beam on the shelter until the bombs impacted. The crews found the flak a considerable distraction and only three shelters were hit; the unit would return to hit the other three in the days that followed.

Other F-111s crews found little difficulty from the defences. Lieutenant Dave Giachetti's target was the chemical weapons storage bunkers at Ad Diwaniyah. Four aircraft had already hit some of the bunkers with laser-guided bombs to open up the structures, then he ran in with the second section of four aircraft to deliver a follow-up attack:

> Our first mission was very straightforward. Others had penetrated the bunkers before us; four went in first to attack the bunkers. Then slightly after them our four-ship came in. Each plane tossed four canisters of CBU-89 gaiter mines [containers with large numbers of area-denial mines] at the target.

Leading the three Tornados attacking Al Asad airfield, Ian Travers Smith was another of those who enjoyed a clear run to the target:

I had a few problems with my autopilot so I had to fly the aircraft manually. I was head-down in the cockpit as we turned on the IP [initial point] for the target run, which was almost along the line of the valley. Then I looked up and I couldn't believe my eyes: all of the runway and taxiway lights were on—the entire airfield was lit up. We really had caught them by surprise. I could see my aiming point, no problem at all. We were absolutely spot on; all the symbology was in the right place. Until we started to drop the bombs, I don't remember being shot at. Then, when we were half way across the airfield, I looked around and saw all these flashing white lights. Not until we were about 20 miles away from the target on the way out did it dawn on me that those 'flashing lights' had been the muzzle flashes of guns firing at us.

At Al Asad the purpose of the attack was to cut the taxiways running between the hardened shelters and the runways. For this the Tornado were equipped with the specialized JP.233 runway-denial weapon. Carried in two large canisters under the fuselage, the weapon comprised 60 runway-cratering bomblets and 430 area-denial mines which were released over a period of six seconds. Once each weapon was clear of the aircraft, a small parachute opened behind it to slow its fall. Each cratering bomb weighed 57lb and was roughly the size and shape of a roadmender's pneumatic drill without the bit. When the bomb hit the surface a primary explosive charge punched a circular hole in the concrete, then a secondary warhead was fired through the hole and into the foundation supporting the surface. Detonating in the confined space between the underside of the concrete and the foundation, the secondary warhead produced an underground cavity topped with a layer of weakened concrete and large amounts of debris. If an aircraft ran over undermined area in the course of taxying, taking off or landing, the latter would collapse and probably 'amputate' one or more of the undercarriage legs. To deter teams attempting to repair the damage, the 430 mines dropped with the cratering bombs ended up among the concrete debris scattered on top of the surface. In the hours to follow they would explode at irregular intervals or if they were disturbed in any way.

At Mudaysis airfield the low-level penetration by three Tornados was similarly successful in achieving surprise, and there was no ground fire until the first of the bombs detonated. Flight Lieutenant Ian Long, bringing up the rear of the attack, watched the bombs from one of the Tornados in front detonate across the airfield:

His bombs went off in front of us with a really bright flash. That was a delight to see—it confirmed that I was heading towards my target. It looked as if everything was going according to the plan. I was not aware of any AAA being fired at me as I was running in.

A few seconds later the attack computer in Long's aircraft began dispensing the bomblets from the containers underneath the fuselage:

There were a lot of bright flashes reflected off the ground, as the weapons were fired down from the aircraft. It was incredibly light around the cockpit. As we passed over the target there were a lot of sparkling lights coming from my left and going behind us, and the odd red ball came past me. It was pretty exciting!

At the other two targets attacked by the Tornados, Tallil and Al Taqaddum, it was a different story. There other raiders had struck at the airfields before the Tornados arrived and the defenders were thoroughly alerted (the JP.233 force had to attack last, or the bombs from following aircraft might have set off the area-denial mines laid across the runways and taxiways). Flight Lieutenant Rupert Clark, piloting one of the Tornados attacking Tallil, recalled:

Ahead of us was this dense wall of AAA. Obviously the fire was not aimed, it was just waving across the sky. Some of it was huge stuff which went up like roman candles, slowly and gracefully. Other bits were whacking around. Some of it was red, some was white. As I ran in the boss's aircraft attacked on my left; I saw a carpet of explosions as the weapons came off and exploded. And then Nick Heard's. Then I saw Rickey Corbelli's burners light up in front of me as he accelerated for his attack. I could see what I was going into and I decided I was not going to turn back—I was going to go through that flak and drop my weapons. Really there was no decision to make, all of my mates had gone through and there was nothing else I could do. But actually, consciously doing it was one of the hardest things that I have ever done. It was the most frightening moment of my life. My target was in one of the thickest bits of AAA, so I moved off a bit to the right to avoid the tracer. I felt my weapons come off and saw the flashes behind me as they detonated, then I heard the thuds as the canisters came off. I applied full dry power for the escape and the jet surged forwards: we had just lost five tons of weight and one hell of a lot of drag.

As Clark and the other three Tornados cleared Tallil, a second section of four of these aircraft ran in to attack other parts in the network of taxiways. After each attacking force came away from its target there was the dreadful moment of truth as the leader sought to discover whether any of his planes had fallen victim to the enemy flak and missiles. Wing Commander John Broadbent, leading the Tallil attack, described the painful moments after he asked his crews to check in:

I said, 'Bristol Formation, check'. Back came the replies '2, 3, 4 (pause) 6, 7, 8.' 'Bristol 5, check?' Nothing. Bristol 5 was my old mates Buckers and Paddy [Squadron Leaders Gordon Buckley and Paddy Teakle]. I thought, Jesus—they've gone down. That was really depressing and all the way back across Iraq I tried to come to terms with it. As we crossed the border we began our climb to rendezvous with the tanker. And then we heard a very faint voice on the radio: 'Bristol 5, checking in.' They had had some radio trouble, that was all.

The initial wave of air strikes on Iraq and Kuwait had been a complete success: the Coalition air forces had flown 671 sorties with manned aircraft and every plane had returned safely. So began the air war in the Gulf. The initial attacks knocked out vital parts of the Iraqi air defence command and control system, setting the stage for the Coalition to establish air superiority within a few days, and air supremacy shortly thereafter. The Iraqis were about to learn, in the cruellest possible way, about the massive destruction wrought by modern air power when there are no effective air defences to prevent it.

15

TORNADO SPYPLANES GO TO WAR

Aerial reconnaissance has come a long way since the first jet reconnaissance mission in the summer of 1944 (see Chapter 8). Today it is a multi-faceted business employing aircraft and drones flying over enemy territory at ultra-low or ultra-high altitude, planes standing-off outside the reach of the defences and looking in or listening from there, and satellites orbiting high above the combat zone. During the recent war in the Persian Gulf the Royal Air Force sent its Tornado GR.1A reconnaissance aircraft into action for the first time. These state-of-the-art planes carry no conventional optical film cameras; instead, they use an electro-optical system similar in concept to the family camcorder to record the scene passing below the aircraft. Photographs are no longer the main product of aerial reconnaissance—now it is 'electro-optical imagery'.

IN JANUARY 1991, a few days before the start of the aerial onslaught against Iraq, six Tornado GR.1A aircraft and nine crews drawn from Nos 2 and 13 Squadrons joined the Royal Air Force Tornado detachment at Dhahran in Saudi Arabia. The GR.1A is optimized for the low-altitude reconnaissance role flying at night or in bad weather, and it carries no optical cameras or conventional film. In the space that had been occupied by the cannon and ammunition magazines in the attack version of the Tornado, the GR.1A carries an in-built electro-optical reconnaissance system. The main sensor is the Vinten 4000 infra-red linescan equipment, which scans from side to side, perpendicular to the line of flight, from horizon to horizon, from a small blister beneath the fuselage. Supplementing this cover, looking to each side of the fuselage, are a pair of British Aerospace/Vinten sideways-looking infra-red sensors. The electronic images seen from these three sensors are fed to six separate video-recorders.

Infra-red photography using conventional film has been around for a long time. Tactically, it has the great advantage that it functions in lighting conditions ranging from bright sunlight to the darkest of nights and requires no artificial illumination (i.e. flares) that would betray the presence of the aircraft. Another well-proven technique is to link the reconnaissance system electronically to the aircraft's navigational computer, so that the latter places in the corner of each image a small block giving the aircraft's position, heading and other details at the time the image was captured; also, as he passes through the target area, the navigator can press a button to put an 'event marker' on any image of particular interest. These features are of considerable assistance to the interpreters who will later examine the imagery. In the Tornado reconnaissance system these features are incorporated and their capability is enhanced.

While the imagery produced by the infra-red electro-optical equipment lacks the crystal sharpness produced by conventional film cameras under optimum conditions, for military intelligence purposes this is a small handicap. The important advantage of the new system compared with normal photography is the reduction in the delay in getting the intelligence to those who need to use it. There is no need to develop or print the imagery before it is viewed. In the aircraft the navigator can observe the video imagery on a television screen in his cockpit in real time (and at night the screen will show things that his eyes may not see), and he can pass on, by radio, any significant discoveries that may have been made. He can even replay in flight particular parts of the imagery if he wishes to identify specific objects on the ground. After the aircraft has landed, the video cassettes can be played immediately for analysis.

During the Gulf conflict the Tornado GR.1As operated as part of a multi-faceted Coalition reconnaissance effort that included several types of drone, F-14s carrying reconnaissance pods, RF-4C Phantoms and Lockheed TR-1s and U-2s. Ground surveillance was carried out by Boeing E-8A (J-STARS) aircraft using a powerful sideways-looking radar to detect traffic movements deep in enemy territory. Electronic reconnaissance (elint) was the domain of the Boeing RC-135 'Rivet Joint', the Lockheed EP-3E 'Aries' Orion and the BAe Nimrod R.1. Overseeing the area at regular intervals were the US satellites with their own secret range of reconnaissance sensors.

Each separate system—the low- and the high-flying aircraft and drones, the radar surveillance planes, the electronic eavesdroppers and the satellites—possessed its own unique advantages for intelligence-gathering. That of the Tornado GR.1A was the ability to conduct searches of specified areas or routes at relatively short notice, and to do so at night and beneath a solid layer of low cloud (which would preclude effective optical or infra-red searches by higher-flying systems).

To avoid optically aimed anti-aircraft fire, the GR.1As operated only at night. Flying singly over enemy territory, these aircraft normally cruised at speeds around 645mph using their terrain-following radar to maintain a constant altitude of 200ft. Although the aircraft had provision to carry a couple of AIM-9L Sidewinder missiles for self-protection, the threat from Iraqi fighters was considered minimal and crews preferred to leave the missiles off and so avoid their weight and drag penalty.

The Tornado GR.1As flew their first combat mission on the third night of the war, 18/19 January. Soon after dark three of these aircraft took off from Dhahran to conduct separate searches of areas from which 'Scud' surface-to-surface missiles were being launched against Israel or Saudi Arabia. Squadron Leaders Dick Garwood and John Hill, assigned to search the area to the south of Habbaniyah, completed their mission without incident. When their imagery was examined afterwards it was found to show a 'Scud' launching vehicle in the open. F-15E attack planes were directed to the area but low cloud prevented them from finding the vehicle.

A second wave of GR.1As also took part of the 'Scud-hunting' effort that night. Flight Lieutenants Brian Robinson and Gordon Walker conducted a search in the Wadi al Khirr area. Later analysis of their imagery showed at least two camouflaged sites thought likely to contain 'Scud' support vehicles.

During the night of 19th/20th Flight Lieutenant Mike Stanway and Squadron Leader Roger Bennett had a brief tussle with the defences. Their mission was a search of the western end of the main Baghdad–Ar Rutbah highway, an area from which 'Scud' missiles were being launched against Israel. Stanway flew along the highway using the aircraft's moving-map display to follow the line of the road, which, apart from the headlights of an occasional vehicle, remained unseen in the darkness. The search continued without incident until the aircraft was some 20 miles east of Ar Rutbah, then, as Bennett later explained, the mission took on a more exciting turn:

I suddenly noticed a bright glow over my left shoulder in my 8 o'clock. I thought it was an IR guided missile, either one of the shoulder-launched variety or an SA-9, and it was guiding towards us on a disconcertingly constant bearing. Mike broke hard left and climbed into it to evade. I selected flares, but the dispenser was faulty and they refused to eject. Fortunately the evasive manoeuvre by itself was enough: the missile went sailing past us and detonated some way away.

Subsequent examination of the imagery revealed a 'Scud' launching bunker with a man standing outside it. It was clear that the man or someone near to him had fired the SAM because the imagery showed that almost immediately afterwards Stanway had banked the aircraft sharply to avoid the upcoming missile.

Invariably it was the highly skilled photo-interpreters (PIs), viewing the expanded imagery on large TV screens in the Reconnaissance

Intelligence Centre at Dhahran, that made all the important Scud finds rather than the aircraft navigators. As Bennett explained,

> One of the PIs found the camouflaged bunker. Once he had pointed out what it was, it was almost obvious. But it required an expert to do it. Everybody tried to find the 'Scuds', but they were not left out into the open waiting to be found. After each firing, the vehicles dispersed and ran back under cover.

Mike Stanway and Roger Bennett had their most memorable sortie during the small hours of 26 February, two days after the start of the Coalition ground offensive. They took off as an airborne reserve in support of two other Tornados that had been allocated specific tasks, but on the way they received orders to fly a route reconnaissance along the main roads linking An Nasiriyah, Al Amarah, Basrah and Jalibah in eastern Iraq.

First the crew had to rendezvous with a Victor tanker over northern Saudi Arabia in order to take on fuel, and that proved no easy task. Thunderstorms in the area caused considerable turbulence, with dense cloud extending from am altitude of 26,000ft down to below 3,000ft. Bennett recalled:

> Normally we would tank at around 10,000 feet. The Victor tanker had tried every level, and at 3,000 feet he was still in cloud and in turbulence. We found the tanker by using our attack radar as an AI [airborne interception radar]. Visibility was down to about 100 metres, with thunderstorms and lightning, and we tanked at 3,000 feet. There was a lot of turbulence, the tanker was moving violently up and down and there was a serious risk of mid-air collision. The weather was awful and getting worse. After a struggle Mike got the probe into the basket but it immediately fell out; he got it back in again and we started to fill up but then the probe fell out again. I looked at the fuel and said, 'Right, we've got enough.' We left the tanker with about 7½ tons of fuel, climbed out the top of the weather at 26,000 feet and headed off to the north.

Just short of the Iraqi border, the Tornado let down to low altitude and headed for Tallil. Bennett continued:

> Still the weather was pretty awful. We did not break cloud until we were below 1,000 feet. At 200 feet we were in the clear, with a solid overcast and no turbulence at that level—perfect conditions in which to do a reconnaissance in a GR.1A!

Near Tallil an SA-8 missile control radar locked on to the aircraft. Stanway hauled the Tornado into a tight turn and Bennett released chaff, and the lock-on ceased. Despite the two pairs of wide-open eyes quartering the sky around the aircraft, no missile was seen and it is likely that none was fired.

The initial part of the reconnaissance, of the highway from An Nasiriyah to Al Amarah, revealed little traffic. Just short of Al Amarah the aircraft turned south and followed the highway to Basrah. The crew saw a moderate amount of traffic, most of it heading north:

> As we ran along that road we were fired at by AAA but fortunately it was not tracking fire—it was unaimed. It looked as if they were firing at our engine noise, and at 560 knots [645mph] at 200 feet they did not hear us until we had gone past. So all of the tracer went behind us—it looked quite pretty!

Short of Basrah the crew turned again, this time on to a westerly heading to follow the highway to An Nasiriyah:

> The road to An Nasiriyah, part of the main Basrah to Baghdad highway, was chockablock with traffic. It looked like the M5 during the rush hour. I didn't need to look at the imagery: we could see the vehicles out of the canopy. They had their lights on, and as we approached they heard us and the lights went out. They probably thought they were about to be bombed. There were all types of military vehicle, including transporters with tanks, all moving west about five yards apart. They were not going very fast, about 10mph. The whole time we were looking out for SAMs, but none came up at us.

Later the crew learned that they had stumbled upon the start of the Iraqi massed withdrawal from Kuwait, later termed 'The Mother of all Retreats', ordered by President Saddam Hussein earlier that morning. Bennett reported the findings by radio to the AWACS aircraft monitoring activity in the area. The Tornado followed the highway for some 60 miles without reaching the head of the column then, its task complete, it turned south and headed for base. The crew had spent more than an hour over Iraq, all of it at low altitude. Dawn was breaking as Stanway and Bennett left enemy territory and they made the final 40 minutes of the flight in daylight. It was the only daylight operational flying time they logged during the entire war.

The build-up of Iraqi traffic was also observed by the Boeing E-8 J-STARS radar aircraft over Saudi Arabia, and several flights of B-52 bombers were diverted to attack the concentrations of vehicles.

During the Gulf conflict the Tornado GR.1A force flew 125 reconnaissance missions, the great majority of which were designated 'successful'. Like most types of intelligence-gathering operation, these carried none of the panache and spectacle associated with the more aggressive types of air operation. Nevertheless, in determining the positions of worthwhile targets, the reconnaissance planes significantly increased the effectiveness of the Coalition attack aircraft.

FINALE

FOR THIS BOOK the author has selected fifteen noteworthy air actions intended to illustrate the truly divers nature of aerial warfare. During the eight decades since its inception, the changes in military aviation have been far-reaching in the extreme. Indeed, military aviation has itself been a major force in driving the advances made in several areas of technology. Yet it has never been enough merely to produce clever pieces of military equipment that operate close to the limits of what is technically possible. Such equipment would be of little avail were it not for the daring, the perseverance and the ingenuity of those who fly in action. As the accounts have shown, those human qualities have never been in short supply.

GLOSSARY

AAA	Anti-aircraft artillery.
AI	Airborne Interception (radar).
ASV	Air-to-Surface Vessel (radar).
Chaff	Metallized strips released from aircraft to create false targets on enemy radar.
Elint	Electronic intelligence.
EOGB	Electro-Optically Guided Bomb.
Ferret	Aircraft equipped for elint missions.
Fritz-X	German guided bomb designed for use against armoured warships.
GCI	Ground Controlled Interception radar.
Geschwader	World War II *Luftwaffe* flying unit with an established strength of about 96 aircraft.
Gruppe	*Luftwaffe* flying unit with an established strength of about 30 aircraft.
HAS	Hardened aircraft shelter.
H2S	Generic term for microwave ground-mapping radar fitted to RAF bombers, produced in several versions between 1942 and the early 1960s.
HARM	High-speed Anti-Radiation Missile. US missile designed to home-in on the emissions from enemy fire-control radars.
Infra-red linescan	Reconnaissance system carried by Tornado GR.1A and other aircraft, to produce electro-optical imagery of targets on video tape for later analysis.
'Johnny Walker'	Code-name for 500lb underwater 'walking' mine, an anti-shipping weapon used by the Royal Air Force.
JP.233	Airfield-denial weapon carried by Royal Air Force Tornado GR.1 aircraft, with runway cratering munitions and anti-personnel mines to delay repair work.
LGB	Laser-Guided Bomb.
'Mandrel'	Radar-jamming equipment to counter the German World War II *Freya*, *Mammut* and *Wassermann* early-warning equipments.
Mark 24 Mine	First-generation air-dropped anti-submarine homing torpedo.
PI	Photographic interpreter.
PR	Photographic reconnaissance.
RHWR	Radar homing and warning receiver.
SAM	Surface-to-air missile.

SD-2	German fragmentation bomb weighing just under 4lb, dropped in large numbers during attacks on aircraft on the ground and on other 'soft' targets. Used for the first time and in large numbers during the initial stages of the attack on the Soviet Union.
Shrike	Early type of US anti-radiation missile used in the Vietnam War, designed to home-in on the emissions from enemy fire-control radars.
Sidewinder	US-designed infra-red homing air-to-air missile.
Sparrow	US-designed semi-active radar homing air-to-air missile.
Standard ARM	Standard Anti-Radiation Missile; US long-range weapon used in the Vietnam War, designed to home-in on the emissions from enemy fire-control radars.
TALD	Tactical Air Launched Decoy. An unpowered radar decoy carried by US Navy planes released from high altitude, whereupon the wings unfolded and it flew a pre-programmed track into enemy territory to lure the defences into action.
'Tallboy'	Code-name for British 12,000lb bomb.
TFR	Terrain-following radar.
'Window'	British wartime code-name for chaff.
'Window Spoof'	Feint operation in which a few aircraft dropping large amounts of chaff give an appearance on enemy radar similar to that from a large force of attacking bombers.
WSO	Weapon Systems Officer (crewman in F-111F and other US Air Force combat aircraft types).

INDEX

Index 175

Smith, Flight Lieutenant E., 39

Smith, Flight Lieutenant N., 56–7

Sollum, 21

Sommer, *Leutnant* Erich, 93–6

South Dakota, USS, 88

Southampton, 40

Spartan, HMS, 64

Specht, *Oberstleutnant* Günther, 122

Spruance, Admiral Raymond, 85–6

Stanway, Flight Lieutenant Mike, 164–6

Stephenson, Pilot Officer Paddy, 35–6

Stockley, Flight Sergeant, 112, 116

Stone, Sergeant Kenneth, 110–11

Strasser, *Fregattenkapitän* Peter, 16

Stryj, 47

Stuka, *see* Junkers Ju 87

Supermarine Spitfire, 29ff, 119, 121

Tactical Air Launched Decoy (TALD), 151–2

Taiho, 87, 89

Tait, Wing Commander J., 102, 106

'Tallboy' bomb, 100, 102, 103–7

Tallil, 160, 165

Target Support, 109, 110, 115

Task Force 58, 82ff

Tate, Captain Steve, 155

Tawi Tawi, 83

Taylor, Flying Officer Pete, 145

Teakle, Squadron Leader Paddy, 161

Tinian, 83

Tirpitz, 98–107

Todd, Squadron Leader Martin, 141–2

Togo, Admiral, 83

Tokyo, 83

Tomahawk cruise missile, *see* AGM-86

Tonkin, Gulf of, 125

Torneträsk, Lake, 105

Toyoda, Admiral Soemu, 83, 85–6

Travers Smith, Wing Commander Ian, 157, 158

Trenkle, *Feldwebel* Fritz, 61

Tromsø Fjord, 104

Tupolev SB-2, 46, 49–50

Tuxford, Squadron Leader Bob, 142, 143, 146, 147

U-boats: *U226*, 55; *U456*, 54; *U657*, 55; *U905*, 57; *U954*, 55

U Tapao, 124

Ubon, 124, 125, 136

Udorn, 124–5

Uganda, HMS, 62

'Ultra', 54, 56, 96

Ungar, *Feldwebel* Friedrich, 74

United States (Army) Air Force(s) units: 1 BD, 68, 69, 73, 77; 1 Combat Wing, 76; 2 BD, 68, 78; 2 BW, 150; 3 BD, 68, 73, 78; 4 BW, 73; 4 FG, 76; 4 FW, 157; 8 TFW, 124, 136; 13 BW, 73, 74, 75; 33 TFW, 156; 45 Combat Wing, 79; 48 TFW, 153, 157–8; 56 FG, 72, 73, 75; 56 SOW, 124; 71 TFS, 155; 78 FG, 72; 91 BG, 76, 81; 94 Combat Wing, 76; 95 BG, 81; 100 BG, 74–5, 81; 352 FG, 120; 353 FG, 72; 354 FG, 76; 357 FG, 79; 365 FG, 120; 366 FG, 120, 121; 381 BG, 76; 388 BG, 79, 80, 81; 388 TFW, 124, 136; 390 BG, 74; 401 BG, 76, 77; 432 TRW, 124, 136; 457 BG, 76, 77

United States Army units: 101 AD, 150

United States Navy units: VB-10, 89; VP-84, 54, 55

Valletta, 26

Van Wagenen, Captain Mike, 134

Vickers-Armstrong: VC-10, 150; Wellington, 109

Victorious, HMS, 99

Volkel, 122

Vu Van Hop, Lieutenant, 128

'*Wacht am Rhein*', *see* 'Watch on the Rhine', Operation

Wadi al Khirr, 164

Wadi Halfa, 22

Walker, Flight Lieutenant Gordon, 164

Wallis, Barnes, 100

'Wandering Annie', *see* 'Mark 24 Mine'

Warspite, HMS, 62

Wasp, USS, 88

Wassermann, 69

'Watch on the Rhine', Operation, 117, 118

Watts, Lieutenant Lowell, 75, 79–80

Weiss, *Hauptmann* Otto, 29

Wideawake, 139ff

'Wild Weasel', 124, 125, 130, 132, 136, 152

'Window', 108, 112ff

'Window Spoof', 109, 110, 115, 131

Wisconsin, USS, 150

Withers, Flight Lieutenant Martin, 140, 142, 143–7

Witts, Wing Commander Jerry, 154

Wizna, 47

Wright, Flight Lieutenant Bob, 144, 147

Wright, Flight Lieutenant J., 54–5

Yap, 86

Yen Vien, 124, 125, 134–6

Yokosuka: D4Y1 'Judy', 84, 86, 88, 89; P1Y1 'Frances', 86

Yorktown, USS, 90

Zemke, Colonel Hub, 75–6

Zuikaku, 90